BUSINESS STRATEGIES FOR THE NEXT-GENERATION NETWORK

BUSINESS STRATEGIES FOR THE NEXT-GENERATION NETWORK

Nigel Seel

Auerbach Publications
Taylor & Francis Group
Boca Raton   New York

Auerbach Publications is an imprint of the
Taylor & Francis Group, an informa business

Auerbach Publications
Taylor & Francis Group
6000 Broken Sound Parkway NW, Suite 300
Boca Raton, FL 33487-2742

© 2007 by Taylor & Francis Group, LLC
Auerbach is an imprint of Taylor & Francis Group, an Informa business

No claim to original U.S. Government works
Printed in the United States of America on acid-free paper
10 9 8 7 6 5 4 3 2 1

International Standard Book Number-10: 0-8493-8035-9 (Hardcover)
International Standard Book Number-13: 978-0-8493-8035-8 (Hardcover)

Library of Congress Cataloging-in-Publication Data

Seel, Nigel.
  Business strategies for the next-generation network / Nigel Seel.
    p. cm. -- (Informa telecoms & media ; 4)
  Includes bibliographical references and index.
  ISBN-13: 978-0-8493-8035-8 (alk. paper)
  ISBN-10: 0-8493-8035-9 (alk. paper)
    1. Computer networks. 2. Business planning. 3. Strategic planning. I. Title.

TK5105.5.S389 2007
004.6--dc22   2006034583

Visit the Taylor & Francis Web site at
http://www.taylorandfrancis.com

and the Auerbach Web site at
http://www.auerbach-publications.com

Contents

Acknowledgments

I would like to thank all of the following whose advice I relied upon implicitly as I put the book together: Sue Davidson for her advice about how NGN carriers can address the SME market; Rob Evans for an insightful discussion of NGN transition issues in an alt-net; David Hilliard, formerly CEO at Mentor and a source of great encouragement; Andy MacLeod, formerly CEO at Verizon Europe, interviewed in chapter 12 and a source of deep insights into the telecom industry and its possible futures; Mike McTighe, formerly CEO of a number of carriers, for his strategic insights into industry evolution; Carol Olney for her authoritative views on the challenges of IT modernization; Bob Partridge, interviewed in chapter 13, for the enormous experience and industry wisdom he shared with me; Stephen Pulman, professor at Oxford University, who is interviewed in chapter 12 and provided an authoritative source on all aspects of natural language processing; Mick Reeve, formerly BT's chief architect, who reviewed a draft of chapter 13 and provided much advice; Alex Seel, who reviewed a number of chapters from the point of view of a telecom programmer (not entirely the intended audience); Dr. Ian Taylor, who reviewed a draft of chapter 11 and whose book on P2P is recommended reading; Yuda Tuval of Mentor for initial encouragement; Andrew Wheen for valuable feedback; Clare Youell for support all the way through, and the figure in chapter 8; and those contributors who have to remain anonymous, all of whom gave so generously of their time.

In regard to the original commissioning of the book and its production, I would like to thank Gavin Whitechurch of Informa, who first suggested that I might write this book; Rich O'Hanley, publisher at Auerbach Publications (part of the Taylor & Francis Group), who commissioned the book and supported me throughout the year-long writing process; Catherine Giacari, project coordinator, who helped me with layout and figures; Marlyn Littman of Nova Southeastern University, for a thorough, helpful review of the first draft, which led to a number of additional topics being included in the final text; Julie Spadaro, Taylor & Francis project editor, for so expeditiously organizing the overall production of the book;

and Lynn Goeller of EvS Communications for handling the copy editing, page layout, indexing and proofreading.

All responsibilities for any omissions or errors are, of course, my own.

Introduction

This is not the first attempt to build the Next-Generation Network (NGN). Back in the 1980s, when the carriers controlled innovation, they had come up with a wonderfully complex architecture for voice, data, and video services, called the Broadband Integrated Services Digital Network (Broadband ISDN). This architecture was layered upon a standard protocol called ATM—Asynchronous Transfer Mode—and those 53-byte cells were deceptively simple. All the real complexity was in the multiple adaptation layers, which allowed very different services to be successfully adapted to and carried by the relatively uncomplicated ATM transport layer, and in the signaling required to make, manage, and tear-down connections.

As we all know, Broadband ISDN took years of preparation, as the standards bodies tried to design in every conceivable requirement before the standard could be finalized and equipment could be built. In the meantime, the Internet happened, using a good enough protocol which couldn't do one tenth the things ATM was supposed to do. But the things it *could* do were what were needed back then, and it was extensible in service.

The current concept of the NGN is emphatically *not* the Internet. The NGN is in reality Broadband ISDN mark 2, leveraging Internet technologies. So is it all going to end in tears again? Hard to say—the NGN specification roadmap is now in the hands of all the usual carrier standards bodies, the ITU-T, ETSI, ANSI, etc., and stretches out past 2009. However, unlike with ATM, the new NGN is leveraging protocols and standards that have some real-world experience behind them, and it's tackling problems of multimedia service networking that we actually have. So it's got to be in with a chance.

Let's assume the new NGN is one of the right answers to the world's networking problems right now (many would disagree, but the premise of this book is that it is near enough). A related question is whether carriers will be able to make the transition from their current networks, processes, and systems to the next-generation network. It will neither be easy nor cheap, and some certainly won't make it. Let me put it like this. Carriers are typically large, complex organizations with poor customer relations and an unusual resistance to change. The next-generation network is a concept and architecture for a complete reconstruction of the way carriers work, based on Internet technologies. Putting the two together, it is obvious—we are going to have a problem.

It is worth reminding ourselves how the Internet came to be. It was certainly not driven by the carriers (although it used their pre-existing transmission and switching networks). The Internet was driven by new-economy vendors like Cisco and a new class of communication companies, the ISPs. We even had a name for the new and old guard: net-heads vs. bell-heads, those who "got it" and those who didn't.

Well, 10 years later carriers have belatedly "got it," or at least the technology part. The Internet is real and its technology base is here to stay. The old carrier dreams of ATM and Broadband ISDN, which they clung to for so long, have finally evaporated. The task now is to re-tool with IP-based platforms. Will the carriers succeed in remaking themselves? Has the Internet merely been a historical transient, a brief period of glasnost before the reimposition of centralized carrier control—business as usual?

When I worked as a carrier architect at Bell-Northern Research, a precursor to Nortel, it seemed to me that our carrier customers had it easy. Carriers had networks, customers, and recurrent revenues. If they did nothing but keep their equipment running, they got paid. By contrast, in Nortel, if we didn't make new sales every day, we didn't get paid at all. We had to struggle for lunch. Many of the people who work at carriers, perhaps even most of them, are not directly involved in the day-to-day operations that keep the cash flowing in. They do things like network planning, product development, marketing, and strategy. Yet at the heart of the carrier is a rigid, process-centric hierarchy: carriers have lots of customers, and serving them all needs a complex machine of processes, people, and IT automation.

Changing this machine is difficult: much easier just to layer the new upon the old, a technique as old as history. When Troy was excavated in modern Turkey, it was found that the site was composed of nine cities layered one upon the other, dating from 3000 B.C. to Roman times. Carriers have their historical layers, too: ancient networks like Telex, asynchronous (PDH) transmission on co-ax or microwave, strange pre-digital voice switches (although most of these are now gone), and X.25 data switches. These are layered below circuit switches, frame relay switches, and the more modern SDH transmission network. Finally, we see the most modern layers such as wave-division multiplexing, IP/MPLS routers, and SIP servers transmuting to IMS call session control function devices in the next-generation network. This forest of acronyms, by the way, is explained in chapter 2.

The 40 or 50 years of network history embodied in the most venerable of our incumbent carriers is paralleled by a similar museum of IT automation: old computers, old programming languages, and old paradigms of computer network architecture. The processes and manual work-arounds that made all this operate end-to-end are still there, and it's just too expensive to modernize them, given that these are legacy products and platforms. It's just that these legacies have real customers with real revenues and the case for keeping them alive seems to win out year after year.

Given the sheer density of distinct roles, processes, automation systems, and ad hoc interfaces needed to keep most carriers in business, the process of transformational change feels like *wading through treacle.*

- Initiatives spawned by senior management get bogged down in the middle management bureaucracy and peter out.
- Expensive programs flow around the edges of the real problems and fail, wasting millions.
- Incremental programs—adding something new—often do succeed, but leave the legacy heartland untouched, and operational costs continue to rise.

Yes, transforming carriers is hard work and attempts at transformation fail far more often than they succeed.

Paradoxically, vendors find change easier than carriers. Lacking recurrent revenue flows, the vendors are more exposed to the volatility of the market—a fact plainly seen after the collapse of the Internet bubble in 2001–02. Market forces smash though the organization and it has no alternative to laying off people, closing some divisions, and reorganizing others. This brutal, creative destruction removes bureaucratization, incompetence and now-redundant activities, and forces modernization. But on the carrier side, even the bankrupt carriers were left operationally intact so they could maintain services to their customers. Clearly superfluous staff *were* laid off, but internal products, processes, and automation were not much changed.

The first part of this book, "Technology," starts with a review of the failure of the previous attempt by carriers to re-tool for the future—Broadband ISDN. I then examine in detail the Next-Generation Network as a set of technologies and capabilities supporting multimedia interactive services, specifically IMS. The third chapter looks in detail at TV delivered over the Internet and Video-on-Demand, but I pay equal attention to business models and changes in the value chain. Finally in this section, I take a look at carrier IT renovation programs.

In the second part of the book, "Transformation," I look at how carriers have attempted to remodel themselves as IP companies. Carriers are perennially trying to move their businesses away from being "mere bit carriers," but we are entitled to ask whether it is always and everywhere such a bad thing to be a bit carrier, and what the alternatives really amount to. The alternatives open to a carrier depend on its position in the marketplace—is it a large generalist, a small niche player, or stuck in the middle? I review some influential thinking about market structure and company strategies.

Business strategies for next-generation networks are about more than technology and marketing. I next examine how to choose the right people and the right roles for transformation projects. It seems obvious that the personal characteristics and skills needed to *drive change* are markedly different from the more operational and routinist aptitudes needed to run a well-oiled (or even badly oiled) machine, but somehow this has been ignored in program after failing program. I finish this section with a "worked example" of how to start up a major change program, and show how personal style can be as important as methodology.

In the third section of the book, "Business and Technology Issues," I identify some more innovative business models. Service Providers such as Vonage and Skype have redefined what voice means for the portfolio, but what has been the carrier response? Proposing to block their traffic has had at least as much air time as more forward-thinking business models and public resources such as Spectrum, which have hitherto been used "free" by carriers, are at last being monetized through such mechanisms as public auctions. A good thing or a bad thing? I then look at Peer-to-Peer Networks, both how they work and the associated politics, economics, and security issues. Finally in this section, I examine the prospects for the automation of natural language understanding and production. You may have had the experience of "talking" with an automated call center agent to book a flight or a hotel. Unless your transaction was extremely conventional and routine, you may have encountered problems that resisted all attempts to "back out." Our next-generation networks are still primarily mechanisms for transporting conversation, yet the networks themselves do not understand what is being said. If this changes over the next few years, what will be the implications?

Finally in the fourth part, we get down to "Business Strategies" in detail. My anchor concept here is that of value nets and market power. Business strategy is fundamentally economic, and is about securing market positions where premium returns can be achieved. In both consumer and business segments, carriers and the NGN are embedded within broader value nets, including content providers and systems integrators. Who wins in this game? I look first at prospects for the incumbents, then at strategies for alternative, competitive network operators, and finally at the consumer market. I conclude that all is not doom and gloom, but the relentless encroachment of commoditization is in fact the back story.

Will the next-generation network mark the reimposition of central control from the carriers, damping down the spirit of freedom and creativity that flowered on the back of the unplanned and unanticipated Internet? I address this fundamental issue in the conclusion.

For additional information and links to relevant resources, go to the Web site associated with this book: http://ngn.interweave-consulting.com/.

Please note that names and dialogue details have been changed throughout, and where roles are mentioned, "he" should be read as "he or she."

About the Author

After a period as a mathematics teacher and commercial programmer, Nigel Seel spent the 1980s in the UK lab of IT&T working on formal methods for software development, artificial intelligence, and distributed computing. He also completed his Ph.D. in artificial intelligence and mathematical logic.

In the 1990s Nigel worked in Bell-Northern Research (Nortel's R&D organization) and later Nortel itself as a carrier network architect, latterly being lead designer for Cable & Wireless's £400 million UK network rebuild in 1998–99. He then freelanced as an independent designer until 2001 when he was hired by Cable & Wireless Global as chief architect, and relocated to Vienna, Virginia. Subsequently he was appointed vice president for portfolio development.

Following the collapse of C&W Global Nigel relocated back to the UK. After more freelance consultancy, he worked with the UK management consultancy Mentor from April 2004 through January 2006. He is currently freelancing again through his company *Interweave Consulting*.

TECHNOLOGY

Chapter 1

The Strange Death of Broadband ISDN

Introduction

In the early nineties I was working for Bell-Northern Research, then Northern Telecom's R&D organization (now Nortel), as a carrier network architect. I recall flying from the UK into the depths of the Canadian winter to attend a Broadband ISDN conference. These days, you may be forgiven for being somewhat hazy as to what Broadband ISDN actually was, but back in the early nineties it was absolutely the next big thing. B-ISDN—Broadband Integrated Services Digital Network—was the universally regarded architecture for the future multimedia carrier network, the Next-Generation Network of its time, underpinned by a packet technology called Asynchronous Transfer Mode (ATM; Bannister, Mather, and Coope 2005). Today, ATM is receding into history, but at the time we all knew it was the future of communications. All traffic: voice, video and data, was to be carried in 48-byte packets, with a 5-byte header for routing and control. This 53-byte packet was called an ATM cell. The magic (and small) numbers of 48 and 53 bytes came about because the main function of ATM, in the carrier view, was to carry voice, minimizing cell-fill latency.

Most data packets, by contrast, are large (e.g., 1,500 bytes) so as to minimize packet header overhead. To carry a large data packet over ATM (Figure 1.1) it is necessary to segment it into small chunks, each of which can be carried in one ATM cell. Extra control information is required to ensure it can be properly put together again at the far end in the right order. The significant extra overhead in segmentation and reassembly, plus the overhead of all those ATM headers, is called the ATM cell tax. IP people in particular deeply resented ATM for this

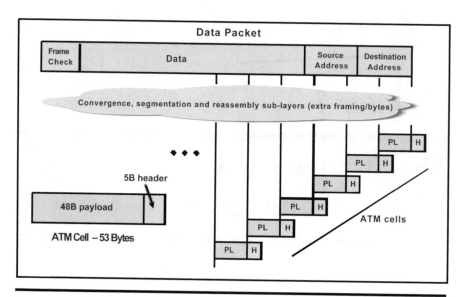

Figure 1.1 Adapting service traffic to ATM.

reason. However, at the time, IP people did not do voice, and so were not especially influential.

Until quite recently, voice was the overwhelmingly dominant traffic on the network, so the networks were designed around it. In a traditional telephone call, the analogue voice signal from the telephone handset is sampled at the local exchange 8,000 times per second and the audio level of each voice sample is encoded in 8 bits (giving 256 possible amplitudes) for an overall 64-kbps signal. In this pre-ATM reality, each voice sample byte is then assigned a "time-slot" and sent across the carrier network to the far-end exchange serving the conversational partner. The network has to be timed to exquisite accuracy so that there is virtually no possibility of losing a voice sample, or of incurring differential delay of successive time slots (jitter). The overall delay (latency) is also minimal, as there is no queuing of timeslots. The circuit-switched telephone network is excellent for basic, vanilla voice. However, all of these desirable properties go by the board when we try to packetize voice in ATM.

In that stuffily-warm, lofty, log-cabin-like conference center in frozen Quebec, leading ATM experts in the field were discussing how to maintain the end-to-end quality of voice calls given that:

- The queuing of ATM cells in ATM switches causes significant cross-network delay (latency).
- Cell queuing times in the ATM switches vary randomly causing jitter.
- Cells are discarded when the queues in the switches get too big, causing clicks and gaps.

I put up my hand and asked why some of the smartest people in the industry were trying to figure out how to put back at the far end of the call the quality of service they had gratuitously thrown away at the near-end (by "cellification" and then ATM network transit). Particularly when we had a perfectly good circuit-switched network, already in service, which simply treated voice properly. I do not recall getting a compelling answer.

My question was both irritating and disingenuous. At the time everyone knew that the circuit-switched network had no future. It had been designed from top-to-bottom to do only one service well: to carry bandwidth-limited voice calls (around 180–3,200 Hz). For reasons later recounted by David Isenberg (1997), trying to get this *one-product network* to do anything else was either impossible or hideously expensive and inefficient. Service innovation was only possible by the carriers themselves, not third parties, and was glacial in pace. No, carrying all traffic in packets or cells was the *only* way to liberate services from hard constraints: if that made it harder to do real-time "isochronous" services like voice, then so be it.

Why Broadband ISDN?

The carriers knew they had to packetize their networks, and that they needed a new architecture (B-ISDN) supported by a number of new *protocols*. Here are some of the functions carriers thought they wanted to support.

- Set up a multimedia video-telephony call between two or more people (involves signaling).
- Carry the call between two or more people (involves media transport).
- Permit access to music, TV programs and varied computer applications.
- Allow computers to communicate efficiently at high speeds.

Each of the above functions would be a chargeable service, leading to billing the customer. Carriers also needed to provision, operate, assure, manage, and monitor their networks as per usual.

When carriers contemplate network transformation or modernisation, they like to huddle amongst themselves in the standards bodies to agree their target services and design a standardized architecture. The latter consists of a number of components, implemented using switches and servers, plus standardized message formats (protocols) to provide the intercommunication. It's easy to see why this ends up as a monolithic and closed activity—it all has to be built by the vendors, slotted together and then work properly. Getting the architecture into service is usually a highly complex and multi-year activity, and the cost will have to be covered by more years of revenue-generating service. Supporters of the model have pointed to the scalability and reliability of modern networks, the accountability, which comes from centralized control, and the sheer functionality that can be put in place by organizations with access to large capital resources and internal expertise.

Critics point to the monopolistic tendencies of capital-intensive industries with increasing returns to scale, the resistance of carriers to innovation and the overall sluggishness and inflexibility of the sector. They note that circuit-switched networking began in the 1860s and that it had taken a further 130 years to automate dialling and digitise calls. By the time I was asking my question in Canada, B-ISDN had already been in gestation for around 15 years with no significant deployment.

The Internet, of course, also took its time to get started. TCP/IP came into service in 1983 and by the late eighties research groups were using e-mail and remote log-in. The fusion of hypermedia and the Internet gave us Web browsers and Web servers in 1993–94 and launched the explosion in general Internet usage. By 1996 there was already a debate within the vendor and carrier community: was the future going to be IP and was the B-ISDN vision dead? It took a further ten years for the industry to completely take on board the affirmative response.

The Internet always ran on carrier networks. More precisely, the basic model of the Internet comprised hosts (computers running an IP stack and owned by end users) and routers (sometimes called gateways) forwarding IP packets to their correct destinations. The routers could be operated by *any* organization (often maverick groups within carriers) and were interconnected using standard carrier leased lines. Almost all hosts connected to the routers by using dial-up modems at each end across switched telephone circuits. So from a carrier perspective, the Internet was simply people buying conventional transport and switched services—the specificity of the Internet was invisible. In truth, the Internet was beneath the radar of the B-ISDN project.

The Internet as the Next-Generation Network

We already mentioned the many complex functions that need to be integrated to make a carrier network work. It's like a highly-specialized car engine. So where was this function for the Internet? Who was doing it? In what is the central mystery of the Internet, no one was doing it. The basic Internet is *unusable*, because it does nothing but provide protocols to allow packetized bits to be transferred between hosts (i.e., computers). It is pure connectivity. However, pure global connectivity means that any connected computer application can be accessed by any other computer on the network. We have the beginnings of a global services platform.

Here are some of the things that were, and are, *needed* to bring global services into being, roughly in the order the problem came up, and was solved.

1. Connecting to a service

Hosts and gateways operate on IP addresses for routing purposes. It is problematic, however, to use IP addresses (and port numbers) as end-system *service identifiers* as well. Apart from the usability issues of having to deal with 64.233.160.4 as the name of a computer hosting a service, IP addresses can also be reassigned to hosts on a regular basis via DHCP or NAT, so lack stability. A way to map symbolic names, such as www.google.com, to an IP address is required. This was achieved by the global distributed directory infrastructure of the Domain Name System, DNS, also dating back to 1983.

2. Interacting with a service

Part of writing an application is to write the user interface. In the early years of computing, this was simply a command line interpreter into which the user typed cryptic codes if he or she could recall them. The introduction of graphical user interfaces in the late eighties made the user interface designer's task considerably more complex but the result was intuitive and user-friendly. The introduction of HTML and the first Internet browsers in the early nineties created a standard client easily used to access arbitrary applications via HTTP across the Internet.

3. Connecting to the Internet

Research labs, businesses, and the military could connect to the Internet in the eighties. But there was little reason for most businesses or residences to connect until the Web brought content and a way to get at it. Initially the existing telephone network was (inefficiently) used for mass connection by the widespread availability of cheap modems. We should not forget the catalysing effects of cheap PCs with dial-up clients and built-in modems at this time. More recently DSL and cable modems have delivered a widely available high-speed data-centric access service.

4. Finding new services

Once the Web got going, search engines were developed to index and rank Web sites. This was the point where Altavista, Yahoo!, and later Google came to prominence.

5. Paying for services

There is no billing infrastructure for the Internet, although there have been a number of attempts to support, for example, micro-payments. In the event, the existing credit card infrastructure was adapted by providers of services such as

Amazon.com. More recently specialist Internet payment organizations such as PayPal have been widely used (96 million accounts at time of writing).

6. Supporting application-application services

Computer applications also need to talk to other applications across the Internet. They do not use browsers. The framework of choice uses XML, and we saw detailed architectures from Microsoft, with .NET, and the Java community with Java EE and companion editions, mostly since 2000.

7. Interactive multimedia services

Interactive multimedia was the hardest issue for the Internet. The reason is that supporting interactive multimedia is a systems problem, and a number of issues have to be simultaneously resolved, as we discuss next. So while for Broadband ISDN, voice/multimedia was the *first* problem, for the Internet, it has also been the *last* (or at least, the *most recent*) problem.

Multimedia Sessions on the Internet

Layering telephone-type functions onto the existing Internet architecture is a challenge. Some of the basics are just not there. For example, the Web uses names asymmetrically. There are a huge number of Web sites out there that can be accessed by anonymous users with browsers. Type in the URL, or use a search engine. Click and go. But the Web site doesn't normally try to find you, and you lack a URL. The Public Switched Telephone Network (PSTN) by contrast names all its endpoints with telephone numbers. A telephone number is mapped to a device such as a mobile phone or a physical line for a fixed telephone. Various companies provide phone number directory services, and the phone itself provides a way to dial and to alert the called user by ringing. The basic Internet structure of routers and computer hosts provides little help in emulating this architecture. Somehow users need to register themselves with some kind of telephony directory on the Internet, and then there has to be some signaling mechanism that can look up the called party in that directory, and place the call. The IETF (Internet Engineering Task Force) has been developing a suitable signaling protocol (SIP—Session Initiation Protocol) since around 1999 and many VoIP companies are using it (Skype is a conspicuous exception, using a distributed peer-to-peer architecture with a proprietary protocol as we discuss in chapter 9).

The next problem is a phone equivalent. A PC can handle sophisticated audio and video, multi-way conferencing, and data sharing. A PC, however, cannot be easily carried in a small pocket. Lightweight and physically small portable IP hosts are likely to have only a subset of a PC's multimedia capabilities and cannot know in advance the capabilities of the called party's terminal—more problems for the signaling protocol. A further reason for the relative immaturity of interactive multimedia services is the lack of wide-coverage mobile networks and terminals that are optimized for IP and permit Internet access. The further diffusion of WiFi, WiMAX and possibly lower charges on 3G cellular networks will hopefully resolve this over the next few years.

Can the Internet, and IP networks in general, really be trusted to carry high-quality isochronous traffic (real-time interactive audio-video)? Whole books have been written on the topic (Crowcroft, Handley, and Wakeman 1999) and it remains contentious. My own view is as follows. In the access part of the network, where bandwidth is constrained and there are a relatively small number of flows, some of which may be high-bandwidth (e.g., movie downloads), some form of class of service prioritisation and call admission control will be necessary. In the network itself, traffic is already sufficiently aggregated so that statistical effects normalise the traffic load even at the carrier's Provider Edge router. With proper traffic engineering, Quality of Service (QoS) is automatically assured and complex, expensive bandwidth management schemes are not required. As traffic continues to grow, this situation will get better, not worse due to the law of large numbers. Many carriers, implementing architectures such as IMS (IP Multimedia Subsystem), take a different view today and are busy specifying and implementing complex per session resource reservation schemes and bandwidth management functions, as they historically did in the PSTN. My belief is that by saddling themselves with needless cost and complexity that fails to scale, they will succeed only in securing for themselves a competitive disadvantage. This point applies regardless whether, for commercial reasons, the carriers introduce and rigidly enforce service classes on their networks or not—the services classes will inherently be aggregated and will not require per-flow bandwidth management in the core.

After establishing a high-quality multimedia session, the next issue of concern is how secure that call is likely to be. By default, phone calls have never been intrinsically secure as the ease of wiretaps (legal interception) demonstrates. Most people's lack of concern about this is based upon the physical security of the phone company's equipment, and the difficulties of hacking into it from dumb or closed end-systems like phones. One of the most striking characteristics of the Internet is that it permits open access in principle from any host to any other host. This means that security has to be *explicitly* layered onto a service. Most people are familiar with secure browser access to Web sites (HTTPS) using an embedded protocol in the browser and the Web server (SSL—Secure Sockets Layer) which happens entirely automatically from the point of view of a user. Deploying a symmetric security protocol (e.g., IPsec) between IP-phones for interactive multimedia has been more challenging, and arguably we are not quite there yet. IMS implements hop-by-hop encryption, partially to allow for lawful interception. Most VoIP today is not encrypted—again, Skype is a notable exception. As I observe in chapter

9, Skype looked for a while to be proof against third-party eavesdropping, but following the eBay acquisition, I would not bet on it now.

Architecture vs. Components

The Internet was put together by many people and organizations, loosely coupled through standard protocols developed by the IETF. Some of it works well, some Internet services are beta or worse. The world of the Internet is exploratory, incremental, and sometimes revolutionary and it's an open environment where anyone can play and innovate. The libertarian ideology associated with the IETF theorizes this phenomenon. The IETF saw (and sees) itself as producing enabling technologies, not closed solutions. Each enabling technology—security protocols, signaling protocols, new transport protocols—is intended to open the door for new kinds of applications. To date, this is exactly what has occurred.

The Internet model is disaggregated—the opposite of vertically integrated. Because the Internet is globally accessible and presents support for an ever-increasing set of protocols (equating to capabilities), anyone with a new service concept can write applications, distribute a free client (if a standard browser will not do), and attempt to secure a revenue stream. This creates a huge dilemma for carriers. In the Internet model, they are *infrastructure providers*, providing ubiquitous IP connectivity. In the classic tee-shirt slogan *"IP over everything,"* the carriers are meant to be the *"everything."* But *"everything"* here is restricted to physical fiber and optical networking in the network core; copper, coax, and radio in the access network; plus an overlay of routing/forwarding and allied services such as DNS. When it comes to end-user services, whether ISP services such as e-mail and hosting; session services such as interactive multimedia, instant messaging, file transfer; or E-business services such as Amazon, eBay, e-Banking, there is no special role allocated for carriers—the Internet model says anyone can play.

This thought is entirely alien to the carriers, who have long believed they were more in the services business than mere bit transporters. Carriers have always wanted to move "up the value chain" whether they were offering network-hosted value-added services or integrated solutions to their enterprise customers. As the carriers came to terms with the success of the Internet, and the collapse of Broadband ISDN, they attempted their own theorisation of the Internet. Not in the spirit of the libertarian open model of the IETF, but more akin to the vertically-integrated and closed models they were used to. They proposed to integrate

- Data and media transport
- Interactive multimedia session management
- Computer application support

into one architecture where everything could be prespecified and would be guaranteed to work. And so arrived the successor to Broadband ISDN, the Next-Generation Network (NGN).

The advantages of the NGN, as the carriers see it, include a well-integrated set of services that their customers will find easy to use, and a billing model that keeps their businesses alive. The disadvantage, as their critics see it, is the reappropriation of the Internet by carriers, followed by the fixing-in-concrete of a ten-year roadmap for the global Internet. The predictable consequence, they believe, will be the stifling of creativity and innovation, especially if the carriers use their NGN architecture anti-competitively, squashing third-party Service Providers, which is technically all too possible.

We should be clear here: anyone offering an Internet service has to develop a service architecture. In the IETF's view of the world, it is precisely the role of Service Providers to pick and choose from the IETF's set of protocol components and to innovate architecturally. There is absolutely no reason why the carriers shouldn't do their architecture on a grand scale through the NGN project if they wish. Critics may believe it's overcomplicated, non-scalable, and ridiculously slow-to-market. If they are right, Service Providers with lighter-weight and nimbler service architectures will win in the marketplace, and the all-embracing NGN initiative will fail. "Let the market decide" is the right slogan, but the market must first of all exist, which means that the Internet's open architecture must be preserved and not be closed down. Many carriers have significant market power and might be tempted to use it in order to preserve what they take to be their NGN lifeline against effective competition, so this is an issue for both customers and regulators. Thankfully, there are reasons to be hopeful as well as fearful.

Why Did Broadband-ISDN Really Die?

There are positive reasons for the smooth uptake of IP, such as the easy availability of the TCP/IP stack as compared with competing proprietary data protocols, the relative simplicity of the basic Internet architecture, and the prior existence of enterprise multiprotocol routers that could be used directly as Internet routers.

Perhaps more important, though, were the problems with ATM. In the 1980s when ATM was being designed, the dominant usage mode was seen to be the multimedia successor to the phone call—human beings making videophone calls. As we saw above, interactive multimedia is the most challenging application for packet networks, requiring a complex infrastructure of signaling, terminal capability negotiation, and QoS-aware media transport. It was not until the mid-nineties that large ATM switches capable of supporting the required signaling and media adaptation came to market—too expensive and too late.

Even worse, the presumed videophone usage model for B-ISDN was highly connection-oriented, assuming relatively long holding times per call. So ATM was designed as a connection-oriented protocol with substantial call set-up and tear-down signaling required *for every call* to reserve resources and establish QoS across the network. This required per-call state to be held in each of the transit network switches. For comparison, millions of concurrent calls (sessions or flows) transit a modern Internet core router and that router knows nothing whatsoever about them.

It turned out that critical enabling technologies for the Internet, such as DNS, require brief, time-critical interactions for which a connection-oriented protocol is inappropriate. Even for connection-oriented applications such as file transfer, which use TCP to manage the connection, connection state is held only in the end systems, not in the network routers, which operate in a connectionless fashion. This has allowed the Internet to scale.

So in summary, ATM had too narrow a model for how end-systems would network, and backed the wrong connection-oriented solution that couldn't scale. Because ATM was designed against a very sophisticated set of anticipated, predicted requirements, it was very complex, which led to equipment delays, expense, and difficulty in getting it to work. The world moved in a direction not anticipated by the framers of B-ISDN and it was stranded, and then discarded.

References

Bannister, J., Mather, P., and Coope, S. 2005. *Convergence technologies for 3G networks,* chap. 7 New York: Wiley.

Crowcroft, J., Handley, M., and Wakeman, I. 1999. *Internetworking multimedia.* Philadelphia: Taylor & Francis.

Isenberg, D. 1997. The rise of the stupid network. *Computer Telephony* (August): 16–26. Also available at: http://www.hyperorg.com/misc/stupidnet.html.

Chapter 2

The Next-Generation Network and IMS

Introduction

In this chapter I begin by looking in some detail at how carrier networks are currently structured and organized. Most of today's carriers are still predominantly voice-centric, with a technology stack supporting circuit-switched voice. Non-voice services such as Frame Relay, ATM, and IP-based services are provided by extra networks, known as *overlay networks*. This is clearly expensive and duplicative. One of the many motivations for moving to the Next-Generation Network (NGN) is its potential to collapse these disparate network platforms into one standardized network infrastructure onto which many different products and services can be easily layered.

I next look in some detail at the NGN as conceived by the carriers today, and as defined in the efforts of the global standards bodies working on next-generation network architecture. At the coarsest granularity, the NGN is structured as three layers: a transport platform (IP/MPLS—Internet Protocol/ Multi-Protocol Label Switching), a session platform (IMS—IP Multimedia Subsystem), and an application infrastructure platform (.NET and Java EE middleware—Java Platform, Enterprise Edition). We will examine each of these layers in turn.

The most important enablers for managed services in the NGN will be IMS, supporting communication services (voice, video-telephony, and many other session-based services) and the IP/MPLS transport platform supporting virtual private network connectivity services. I address both of these in more detail in Appendix 1 and 2 at the end of the chapter.

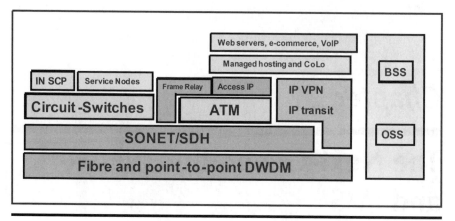

Figure 2.1 The current-generation network architecture.

The Current-Generation Network

In carrier networks to date, the major division has been between *circuit switching* and *transmission* (transmission divides into SONET/SDH—Synchronous Optical Network/Synchronous Digital Hierarchy and optical networking using DWDM—Dense Wave-Division Multiplexing) (Stern, Bala, and Ellinas 1999). Traditionally, both switching and transmission have been voice oriented (Figure 2.1).

The switching/transmission divide is not just technological, but also a structural feature of organizations and even engineering careers. There are still many telecoms engineers around who will proudly state they are in switching or transmission, and each will have a less-than-detailed view of what the other discipline is all about. Data people, formerly X.25, Frame Relay, and ATM, and latterly IP, were historically the new, and rather exotic next-door neighbors.

Circuit Switching

The traditional problem of *switching* is essentially one of connection: how to identify end-points (by assigning phone numbers), how to request a connection between end-points (by dialing and signaling) and how to physically set-up and tear-down the required voice connection (using telephone switches). Once upon a time this was done by analogue technologies, but that is going back too far. From the 1980s, the state of the art was digital switching, and telecom voice switches became expensive versions of computers.

Once people were digitally connected, more advanced services could be introduced such as free-phone numbers, premium rate numbers, call blocking, call redirect, and so forth. Initially this was done by increasing the complexity of the call-control software in the digital telephones switches. Unfortunately, such code was proprietary to the switch vendors: the carriers paid handsomely to buy it, and were then locked-in for their pains. The solution was for the carriers to get together and design a standardized architecture for value-added voice services called the Intelligent Network (IN). In North America the preferred term was Advanced Intelligent Network (AIN). The IN architecture called for relatively dumb switches (service switching points—SSPs) invoking service-specific applications running on high-specification computers called service control points (SCPs) during the progression of the call. Since the very same vendors sold SSPs and SCPs as sold the original switches, prices did not go down and the IN was only a partial success at best.

Transmission

Transmission solves a different problem—that of simultaneously carrying the bit-streams corresponding to many different voice calls, data sessions, or signaling messages over long distances on scarce resources, such as copper wire, coaxial cable, radio links, fiber optic strands, or precisely-tuned laser wavelengths. A transmission engineer would start with a collection of nodes—towns and cities where telecoms equipment was going to be placed—and an estimated traffic matrix showing the maximum number of calls to be carried between any two nodes. The next step was to design a hierarchy of collector links that aggregated traffic from smaller nodes to larger hub nodes. These hubs would then be connected by high-capacity backbone links. This sort of hub-and-spoke architecture is common in physical transportation systems as well: roads, rail, and air travel.

Voice traffic never traveled *end-to-end* across the transmission network, because it had to be routed at intermediate voice switches. The telephone handset connected to a local exchange switch (or a similar device called a concentrator) at a carrier Point-of-Presence (PoP) located within a few miles of the telephone. The local switch or concentrator then connected to transmission devices to send the call to a much bigger switch at the nearest hub. From there, the call bounced via transmission links from switch to switch until it reached the called telephone at the far end.

Switch engineers called the transmission network "wet string," based on the child's first telephone—two tin cans connected by wet string (wetting decreases sound attenuation). Transmission engineers, on the other hand, considered voice switches as just one user of their transmission network, and in recent years a less interesting user than the high-speed data clients. These are to voice switches as a fire hose is to a dripping tap. For transmission engineers, it's all about speed and they boast that they don't get out of bed for less than STM-4 (Synchronous Transfer Module level 4, running at 622 Mbps).

Just a note on terminology. The word "signal" is used in two very different ways. In session services such as voice and multimedia calls, signaling is used to set up and tear-down the call as previously noted. Here we are talking about a signaling *protocol*. However, in transmission, signals are just the physical form of a *symbol* on the medium. So, for example, we talk about analogue signals, where we mean a voltage waveform on the copper wire copying sound waves from the speaker's mouth. We talk about digital signals when we mean bits emitted from a circuit, suitably encoded onto a communications link (cf., digital signal processing). The two uses of the word "signal" are normally disambiguated by context.

The Next-Generation Network

The Internet itself is a platform that today embodies *some* next-generation network technologies. It is a major thesis of this book that over the next five years or so, carrier NGN investment programs will recreate the Internet as a next-generation network, but one where the carriers are full players rather than the somewhat bemused bystanders that they are at the moment. However, a number of research networks already exist that anticipate some of what is to come, a few are described in Table 2.1.

Looking at these networks alone will give a misleading impression of the NGN, however. These are *research* networks, designed to deliver a high-quality, high-bandwidth service to research and educational institutions and to sponsor

Table 2.1   Some NGN Initiatives

NGN	Description
GEANT2	• European Commission 10 Gbps network • Links 34 countries across Europe • IPv4, IPv6 and CoS support
GRNET2	• Greek University and Schools network linked to GEANT2 • Focus on Grid Computing • Business, training and e-learning application
Internet2	• A universities-industry consortium for high-speed networking • Created the Abilene network, close links with NLR (next) • 10 Gbps with 100 Mbps per user
National LambdaRail (NLR)	• 10 Gbps IP network • Gigabit national switched Ethernet service • 40 wavelength DWDM
SURFnet6	• University, Medical and Research network in the Netherlands • Hybrid packet and optical services network • Institutions connect at up to 10 Gbps.

innovative applications. They are a classic instantiation of the "stupid network" discussed in chapter 1, with innovation occurring at end points. The NGN, however, is not like that. It is the invention of carriers who see a pressing need to develop revenue-generating service infrastructure on top of their basic connectivity networks. The NGN is a layered construction, with most of the novelty at levels higher than IP or optical transport. This goes to the heart of the NGN controversy. Does it fundamentally break the Internet model that has led to so much creativity and innovation over the last decade? Or is it precisely bringing those very innovations to the mass market, and creating a more advanced platform for the next round of innovation? We will encounter this fundamental dichotomy again and again throughout this book, but first we need to examine in detail the NGN layering model itself.

Many books cover aspects of the architecture and technology of next-generation networks; see Huitema 2000; Camarillo and Garcia-Martin 2004. In this section I want to look more at the intent behind the design. The Next-Generation Network exploits the highly-decoupled IP transport architecture of the Internet, and more generally, the modular IETF (Internet Engineering Task Force) and W3C (World-Wide Web Consortium) protocol architectures, to create a more cleanly layered model than the existing PSTN (Figure 2.2).

Transmission Layer

The purpose of the transmission layer is to carry communications traffic across the network. There have always been a diversity of options—the copper from your home to the nearest exchange carries analogue signals from a traditional voice phone. If you have DSL broadband, it also carries digital signals to/from your computer encoded as modulated high-frequency tones. If you have cable, then

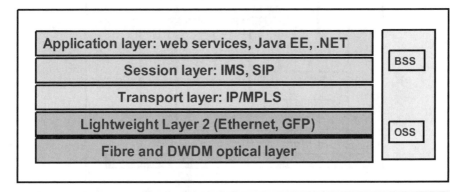

Figure 2.2   The next-generation network architecture.

digital TV content (downlink) and digital data traffic (two-way) are carried on modulated carrier waves across a broad spectrum within the coaxial cable. If you have WiFi, then the same digital signals are carried through the air as modulated radio waves before transitioning to copper DSL or coax.

In the core of the network, torrents of data at 40 gigabits per second are carried via laser light on multiple modulated wavelengths on a single strand of optical fiber. This is Dense Wave-Division Multiplexing (DWDM) contrasting with earlier/cheaper technologies that use fewer wavelengths (Coarse Wave-Division Multiplexing—CWDM).

Most IP networks today will still use SONET/SDH as the transmission protocol encapsulating packet traffic (IP, MPLS) within SONET/SDH virtual containers. The NGN protocol stack (Figure 2.3) comes, in fact, with plenty of options, not all shown in the diagram.

SONET/SDH was originally designed to multiplex thousands of 64 kbps telephone calls very efficiently onto one fiber bearer. In fact SONET and SDH, as protocols, are almost identical, the differences being in the functions of a few header fields, and in terminology. In this book I will mostly use the language of SDH, but with cross-references to the SONET terminology where appropriate. Why did the world not go for one standard rather than two almost identical ones? The place to look is the structure of the telecommunications equipment industry, with large North American companies (Lucent, Nortel) confronting large European companies (Alcatel, Ericsson, Siemens) at the time. Vendors are particularly influential in the standards bodies.

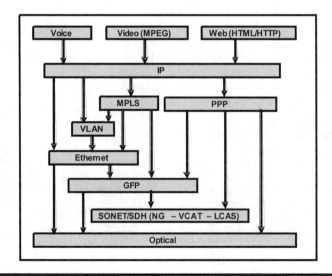

Figure 2.3   Some NGN Protocol Stack Options.

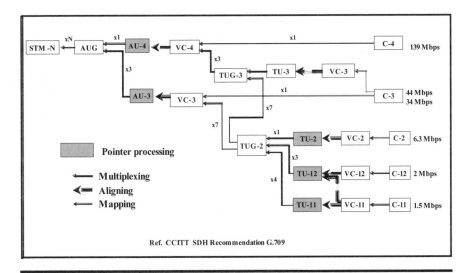

Figure 2.4   SDH Multiplexing Structure.

The lowest rate of SONET/SDH is STM-1 (Synchronous Transfer Module level 1) in SDH, or in North America OC-3 (Optical Carrier level 3). This can carry up to 2,016 64 kbps channels on a 155 Mbps link. Figure 2.4 shows how lower rate "containers" are multiplexed into higher-rate containers to eventually fit into an STM frame. It is not the intention to explain the SONET/SDH multiplex structure here in detail. For compatibility reasons, SONET/SDH containers had to be designed to fit North American DS-1 frame structures carrying 24 time slots at 1.544 Mbps, as well as European E1 frame structures carrying 32 timeslots at 2.048 Mbps. This accounts for the VC-11 and VC-12 virtual containers. Similar motivations prompted the higher rate containers (VC-3, VC-4).

Much faster line rates than STM-1 are possible: STM-4/OC-12 at 622 Mbps, STM-16/OC-48 at 2.5 Gbps, STM-64/OC-192 at 10 Gbps, and recently STM-256/OC-768 at 40 Gbps. STM-1024/OC-3072 at 160 Gbps is still work-in-progress at time of writing. Data packets can also be carried within the SDH protocol structure. Originally each data stream had to fit inside an STM-1 level container, but recent developments aimed at keeping SDH current have seen containers "virtually concatenated" together to form an end-to-end link of much higher bandwidth. This new version of SDH sounds like a TV series—SONET/SDH Next-Generation.

The SONET/SDH-NG "virtually concatenation" facility (VCAT) comes with a protocol (LCAS—Link Capacity Adjustment Scheme) to allow capacity to be dynamically adjusted across the network. SONET/SDH-NG has three further advantages going for it.

- It has a sophisticated operations and management channel allowing the transmission network to be easily configured, monitored, and managed in real-time.

- It has a sophisticated protection architecture: if there is a fiber failure or equipment malfunction, this can be detected and the traffic switched around the problem in less than 50 milliseconds—unnoticeable in a voice conversation.

- As a synchronous network, very precise timing information is distributed all around the network, and can be read off the SONET/SDH line systems. This is not only useful to synchronize all kinds of equipment across the network, it is also used by customers as a service to synchronize and phase-lock their own equipment. Packet networks are not synchronous and do not provide such a service.

All these functions are being replicated in MPLS, but interestingly, the timing issue seems to be the hardest to solve at the time of this writing.

IP/MPLS Transport and Routing Layer

This is the classic Internet model. In an all-IP world, hosts, or end systems (computers, servers, or anything that can run an IP stack) communicate over any convenient wired or wireless access transmission link (wet string) to edge routers. These edge routers look at packet headers and then forward them on the correct "next hop," to the next router in the chain, or to the final destination host. Routers started as ordinary computers running routing software (this still works!) in the earliest days of the Internet, and then became special purpose machines with a custom architecture. Initially focused on enterprise applications, a new generation of ultralarge and ultrareliable machines came into service in the Internet boom of 1999–2001. The current state of the art is the Terabit router, the Tera prefix (10^{12}) indicating aggregate router throughput of thousands of billions of bits per second.

Only routers at the *edge* of modern Service Provider networks actually see IP packets. The Provider Edge routers encapsulate a packet into MPLS by attaching a label to the front of the packet that indicates its final destination (unlike IP addresses, labels are only locally unique and may be altered at each intermediate label-switching router, thereby supporting scalability). Interior or core routers forward the labeled packets—based on their label information—along label-switched paths. The threading of label-switched paths through network routers is under the explicit control of the operator, and this control is used for a number of purposes:

- Load-balancing between alternative routes,
- The creation of virtual private networks (VPNs) for enterprises,
- Segregation of traffic between different quality of service classes,
- Network survivability via failover to backup label-switched paths.

There used to be many concerns about the robustness and service quality of IP networks in general, and the Internet in particular. But as the Internet has become more central to business, significant care and attention, as well as capital resources have been invested by telecom carriers. The Internet is no longer a byword for flakiness and delay. Many carriers privately believe that the Internet is currently "too good," and as the inexorable rise of Internet traffic fills up the currently rather empty pipes, expect to see a harder-nosed commercial attitude emerging. More on this in chapter 13.

Many carriers will focus their NGNs on connectivity services based directly on IP such as Internet access and MPLS-based VPNs (discussed further in Appendix 2). Services such as leased lines, frame relay, and ATM will either be discontinued, or will be supported only on legacy platforms that will eventually be phased out—this may take a while for leased lines services based on SDH, for example. However, some carriers want to phase out and decommission legacy *networks* early, to get the OPEX advantages of a simpler network infrastructure, but still leave these legacy *services* in place to avoid disruption to customers.

Surprisingly, there is a way to do this. It involves emulating leased line, Frame Relay, and ATM services over the new MPLS network, using service adaption at Multi-Service Access Nodes (MSANs) or Provider Edge routers at the edge of the network. There is obviously a considerable cost in MSAN or edge router device complexity, service management overhead and in dealing with the immaturity of the MPLS standards for doing this (using MPLS pseudo-wires). The advantage seems to be in decoupling *platform* evolution to the NGN from *portfolio* evolution to "new wave" products and services.

GMPLS

Following the success of MPLS in the provision, configuration and management of virtual circuits for IP networks, some thought was given as to whether MPLS might be used to handle other sorts of virtual circuits, not as a transport mechanism, but as a signaling and control plane for:

- Layer-2 virtual circuits for Frame Relay,
- TDM virtual paths for SONET and SDH,
- Wavelengths in an optical transport network (OTN),
- Fiber segments linked via spatial physical port switches.

Thus was born Generalized MPLS (GMPLS), which applies the MPLS control plane (with extensions) to these other layers—the focus is typically on SONET/SDH and optical (wavelength) networks. Cross-connects and Add-Drop Multiplexers in these networks need to exchange GMPLS protocol messages. This is not necessarily strange—all these devices today run element managers or SNMP agents that communicate via a management IP layer. In the TDM world, MPLS label allocation/de-allocation is identified with time-slot allocation; in the optical world it is equivalent to wavelength allocation.

In fact, GMPLS has had a mixed reception in the world's carriers. Optical and transmission engineers don't necessarily believe that the IP guys know best when it comes to controlling their networks. There is a history of virtual path management in SONET/SDH networks and optical channel management for OTNs being organized through the network management systems. With increasing element intelligence and more powerful management tools, it is widely felt that the management plane is adequate, and that replicating its functions in a new signaling plane is not required. Of course, opinions differ.

Ethernet Provider Backbone Bridge/Transport (PBB and PBT)

Even MPLS is being challenged. As carriers move to a simplified and more cost-effective technology stack, the prospects of carrying IP within Ethernet directly over the OTN seem increasingly attractive. Traditional Ethernet has scaling and management problems, because its forwarding model depends on Ethernet switches flooding outbound network links with Ethernet frames where the destination is not known, and then learning which exit port to use in future by noting which port the reply eventually arrives at. For this to work, there has to be a unique path between any two points on the network, which is guaranteed by Ethernet's spanning tree protocol. This turns off network links between Ethernet switches until a minimal covering tree remains, but the procedure has a number of problems including inefficient use of network resources and long recovery times in the event of link or node failure (a new tree has to be recomputed and established).

However, using Provider Backbone Transport, a subrange of VLAN tags is reserved for carrier forwarding purposes, the chosen tags (+ destination addresses) functioning somewhat similarly to MPLS labels. This forwarding information is provisioned into the carrier Ethernet switches by the central network management function, or by GMPLS, to create forwarding virtual circuits (and optionally failover restoration paths) across the carrier network. Unlike the situation with MPLS labels, however, the PBT combination of destination MAC header and VLAN tag is globally unique and identical across network switching elements. This offers significant advantages over MPLS in fault finding and tracing.

Carrier Ethernet is seeing a number of innovations that increase its capabilities, including:

- MAC-in-MAC—the provision of a separate carrier forwarding address field, pre-pended to the enterprise customer header (defined in 802.1ah, and also called "Provider Backbone Bridge")
- Q-in-Q—used to create a hierarchy of VLAN tags allowing carriers to distinguish between customers (defined in 802.1ad, and also called "VLAN stacking").

With these, Ethernet is now beginning to match MPLS accomplishments, both in traffic engineering and in the provision of customer VPNs. The battle to come will prove interesting.

IPv6?

Another issue that surfaces on a regular basis is the future of the current version of IP, IP version 4. There has been debate over many years as to when, or whether, the Internet should transition to IPv6. My own position on this is skeptical for the following reasons.

Recall that the Internet is a network of networks. The Internet backbone is composed of networks from tier-1 Internet companies such as Verizon, AT&T, Sprint, British Telecom (BT), Deutsche Telekom, and so on. Smaller tier-2 carriers connect to the tier-1 companies, and in their turn offer connectivity to even smaller tier-3 ISPs. All of these networks are currently running IPv4 for Internet traffic. Since a collective cutover to IPv6 is not on the cards, any protocol migration to IPv6 is fraught with practical difficulties. The early mover will encounter guaranteed IPv4-IPv6 interworking issues, will gain few advantages from the move, and will contemplate wistfully the many advantages that would have been gained from sticking with IPv4 for the duration.

When the IETF designers finalized IPv6 back in 1994, they had added to it many attractive features over IPv4. These included support for end-to-end security via IPsec, support for class-of-service marking via a new Differentiated Services field in the IPv6 header, support for host mobility (mobile IP), support for auto-configuration and, of course, the much larger address field. Unfortunately for IPv6, time has whittled away many, if not all, of these advantages. To put it especially bluntly, most everything of value in IPv6 was re-engineered back into IPv4 on the understandable basis that the world couldn't wait.

- IPv6's class-of-service marking scheme replaced IPv4's obsolete "type of service" header field as the new IPv4 Diffserv Code Point—DSCP.

- IPsec was engineered to work with IPv4.
- Mobile IP was engineered to work with IPv4.
- DHCP configuration of IPv4 hosts obviated most of the IPv6 auto-configuration facilities.
- Private addressing and NAT resolved the address space problem in practice.

Despite claims that these engineering hacks would impact on usability, the difficulties have been steadily overcome. Even some of the hardest problems, getting signaling applications for VoIP to work through NAT and firewalls, have now been mostly solved. Skype is a case in point, following the earlier pioneering work in network gaming and peer-to-peer file sharing.

There is a purist motivation for IPv6, which looks to get back to a clean and transparent end-to-end Internet model leveraging the larger address space of IPv6. NAT is particularly disliked, seen as breaking the simplicity of host-router transparency. However, with NAT making some contribution to network security and working well in practice, the practical motivation is less strong. Given the lack of a positive business case for the IPv4 to IPv6 transition, together with the effectiveness of the IPv4 "workarounds," it is hard to predict whether the transition to IPv6 will ever happen. One positive but seldom-mentioned feature of IPv4 against IPv6 is that 128 bit IPv6 addresses such as:

$$2001:0db8:85a3:08d3:1319:8a2e:0370:7334$$

are a lot harder for humans to manage than IPv4's 66.249.64.4—even after the many rules for presentationally shortening IPv6 addresses have been taken into account.

Session and Application Framework Middleware Layers
The Internet operated for a long time with a few major protocols doing the heavy lifting. TCP (Transmission Control Protocol) provides a reliable connection for moving files, and HTTP (HyperText Transfer Protocol) is used by browsers, running over TCP, to connect with Web servers and retrieve Web content. However, these protocols are of no use for setting up and managing dynamic voice, video, instant messaging, and data sessions between end users.

The Road to IMS
As the early experiments with voice over IP evolved into services with significant usage, it was clear that a multimedia signaling protocol was required, analogous to the signaling used in the existing telephone networks (most notably common channel signaling system 7 often loosely referred to as "SS7"). Multimedia signal-

ing over IP networks was always going to be more complex. User terminals had to negotiate with each other to determine their media-handling capabilities, and with the network to request the quality of service they needed. There were issues of security, and problems in finding the IP addresses of other parties to a call.

The carriers, through the ITU, had an existing protocol suite, H.323, which had been developed for LAN-based video-telephony. This was pressed into service in first-generation VoIP networks, but its clumsiness and lack of scalability triggered activity within the IETF to develop a more IP-friendly, extensible and scalable signaling protocol. Over a period between 1996 and 2002, the IETF developed the Session Initiation Protocol (SIP) as the end-to-end signaling protocol of choice for multimedia sessions over IP. SIP languished for several years, waiting for other developments to catch up, when perhaps surprisingly, the initiative was taken by the cellular industry.

The Third Generation Partnership Project (3GPP) was set up in 1998 to specify a third generation mobile system evolving from GSM network architecture. At the same time the 3GPP2 organization was set up by standards bodies in the US, China, Japan, and South Korea to fast-track a parallel evolution to third generation mobile, evolving from the second generation CDMA networks prevalent in those countries.

3G mobile architecture had originally used ATM, but by 2000 it was clear that the future was IP. Arrangements were therefore made to set up formal links between 3GPP/3GPP2 and the IETF to develop IP standards for 3G. The 3G subsystem that handles signaling was the IP Multimedia Subsystem (IMS) and was based on the IETF's SIP. But in the IMS architecture, SIP had to do a lot more work. For example, users may want to set up preconditions for the call to be made (e.g., QoS or bandwidth) before the called party is alerted, or they may need information on their registration status with the network, and terminals need SIP signaling compression on low-bandwidth radio access links to speed-up transmission and reduce contention for bandwidth.

What the 3GPP communities really needed was an *architecture* that could standardize the interrelationship between the many functions needed to bring 3G multimedia services into commercial reality. Such an architecture would have to integrate many different protocols (signaling, authentication and authorization, security, QoS, policy compliance, application service management, metering and billing, etc.). To deal with the many new developments to SIP (and other protocols) that were needed to make the IMS architecture work, a joint 3GPP-IETF group, SIPPING, was set up.

As the 3G Mobile architecture evolved, it came to the attention of architects and standards people working on evolution for the *fixed* network operators. This Next-Generation Network activity, carried out in bodies such as ETSI TISPAN and the ITU-T NGN program, had a similar requirement for an all-IP signaling layer and session management architecture. IMS essentially fitted the bill, and was adopted, although further changes and developments are in the future roadmap. So, for example, BT's twenty-first century network architecture will eventually have IMS right at the center.

IMS has been described as "mind-numbingly complex." This may be true, but the complexity is there for a reason. IMS provides common services to: user terminals, network-based application servers, network routers, policy engines, billing systems, and foreign networks for roaming capabilities. It provides for authentication, registration and security. It supports presence and instant messaging, and new services such as *Push to Talk over Cellular* (PoC). By doing so much, through standard interfaces, the intent of IMS is to remove the need for new services to re-invent these wheels. IMS-powered services should therefore be lighter-weight and be more easily introduced. Carriers believe they will take a one-time hit to get IMS into their networks, and will afterwards reap the benefits over subsequent service introduction.

At the lower layers of the network, there was little dispute as to who provided the service. Running optical/SDH transmission networks, and running IP networks is pretty much definitional as to what carriers do. But as we get higher in the stack, the focus turns more to applications running on servers that exploit the IP network for connectivity. You don't have to be a carrier to run servers. In principle, a multimedia telephony company could run IMS in a garage. IMS is not precisely designed to do this, because it was conceived by carriers, who arranged for a high degree of potential coupling between the IMS layer and the IP layer. However, this coupling does not have to be turned on, and may not even be necessary for many service concepts, rescuing the garage option. Or perhaps the carriers could be encouraged to expose the necessary IP transport layer interfaces specified in IMS to third parties? And, of course, multimedia telephony companies who do *not* use IMS today (e.g., Vonage, Skype) do indeed build their businesses on servers (rather few in Skype's case) and then buy-in the Internet connectivity they need.

A generalized, powerful, and complex session management platform such as IMS is rather pervasive and there may be a case for it being provided by a specialized ISP, or a facilities-based carrier (a carrier that owns telecommunications equipment—normally fiber, transmission equipment, routers and switches). However, when it comes to providing a discrete service such as music download, access to streaming audio or video, or any specific application service, there is little reason to believe that facilities-based carriers have some special advantage. Most of us don't do our Internet banking with the company providing our broadband connection. This should be a warning to carriers not to go too far down the "walled garden for content or value-added services" path as the route to future margin success.

Application Platforms

At the most general level, a network can be conceptualized as a mechanism, frequently drawn as a cloud, connecting any subset of (people, data, applications) together. SIP/IMS is optimized for connections involving *people* to people and data, because the session holding times are typically long and the user-interface properties have to suit people (audio, video adapted to the terminal device capabilities).

When applications connect to applications, they exchange formatted byte streams. The sessions can be ultrashort, and protocol support is required to choreograph sessions, as there is no human common sense to rely upon. This is the world of computer record exchange, remote procedure calls, asynchronous communications, transaction capabilities, and session management protocols. XML has emerged as the syntactic framework of choice for application internetworking, and both Microsoft and the Java communities have developed application platform architectures. These platforms are Microsoft's .NET and the Java community's Java EE (Java Platform, Enterprise Edition) respectively. They run on computer servers connected to the Internet and provide a preexisting platform onto which E-business applications can be installed. Java EE and .NET systems talk across high-quality IP/MPLS transport networks. Carriers get to play by providing a JAVA EE or .NET hosting service so that customers can install their applications via standardized interfaces. The carrier may also provide many other useful services:

■ Data backup and restore,
■ Hosted application development environment,
■ Managed applications, web server, application server,
■ Caching and content distribution services,
■ Security monitoring.

It is fair to say that hosting of application platforms is at the cutting edge of hosting services today. Most carriers are happier providing managed servers, network connectivity, and application monitoring on top of operating systems such as Windows and Unix/Linux.

Fixed-Mobile Convergence

The discussion so far has been mostly focused on carriers with fixed access networks. At some level of abstraction, the current architecture of mobile networks is not too dissimilar to that of the fixed network. Figure 2.5 shows the case for GSM networks.

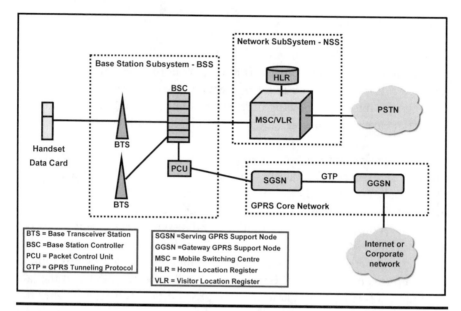

Figure 2.5 GSM Network Architecture (voice and data).

The main differences are obviously the mobile handset and mobile Radio Access Network (RAN). The RAN architecture comprises base transceiver stations (BTS) that handle the wireless link and mobile terminal hand-off, and the base station controllers (BSC) that control BTSs. BTSs and BSCs are collectively named the Base Station Subsystem (BSS). The BSS devices work into the mobile operator's circuit switches (mobile switching centers—MSCs) that are loaded with special software for handling mobile handsets. The MSCs are tightly coupled with special databases/directories: the HLR (Home Location Register) stores subscriber details for registered customers, and the VLR (Visitor Location Register) stores information about handsets registering from "foreign" networks.

The GSM data architecture (GPRS—General Packet Radio Service) uses routers to implement access and gateway functions. The SGSN (Serving GPRS Support Node) is the point of connection for the GSM mobile handset or data card while the GGSN (Gateway GPRS Support Node) controls the interface between the access network and the final destination network (the Internet or an enterprise network, for example). User IP traffic is tunneled between SGSN and GGSN using GTP (GPRS Tunneling Protocol), providing mobility management. As the mobile user moves around, they may be migrated from BSS to BSS, and from SGSN to SGSN—the GTP tunnel end point is moved from SGSN to SGSN, while the allocated GGSN (anchoring the other end of the GTP tunnel) is retained.

Note that the actual management of mobility does not use mobile-IP, but relies upon GSM layer-2 BSS and SGSN handover mechanisms. This understandable design decision has unfortunately propagated through to the European (3GPP) version of IMS, where it will create needless difficulties in integrating IMS with fixed and non-2/3G wireless networks. The North American 3GPP2 community is more fortunate.

Past 3G there is 4G, still in the early research stages. The concept of 4G cellular seems to be centered around much higher data rates (100 Mbps seems to be the target supporting powerful multimedia services), with new modulation schemes and antenna designs. I have not seen any substantive work on systems architecture on the network side of a proposed 4G radio access network. To tell you the truth, I am rather reminded of the early days of ATM, back in the 1980s, when there was speculative work on the "successor to ATM." I believe that it would have relied upon ultrashort cells, and some of the work eventually ended up in AAL-2. But the world didn't go that way, and that is why, in a world with WiFi and WiMAX in it, I am completely agnostic about 4G. For more detailed information on all aspects of 2G and 3G voice and data networking, see Bannister, Mather, and Coope (2004).

Of course, the 2G/3G architectures and operators are not the only wireless game in town. We have already mentioned the short-range WiFi, as found in homes, enterprises, and public "hot-spots." Over the next few years, the Metropolitan Area Network wireless technology known as WiMAX will begin installation. As is well-known, WiMax comes in a fixed DSL-emulation mode (802.16-2004) and a mobile variant that supports the attachment of mobile terminals and inter-cell handoff, called 802.16e. Mobile WiMAX will likely be 18 months later then 802.16-2004 and have a reduced range as compared with fixed WiMAX.

There was a time when a combination of WiMAX MAN and WiFi LAN technologies looked like creating a low-cost mobile Internet that would put the mobile operators out of business. There were visions of fixed operators investing massively in this new technology to grab back the customers they were losing to the mobile sector. Since both WiMAX and WiFi are IP technologies, this would obviously drive VoIP, and since VoIP cannot work without a signaling layer, there would therefore be a knock-on driver for the rapid deployment of SIP solutions, such as (but not restricted to) IMS.

Sadly or otherwise, the arrival of the mobile Internet has been delayed. The fixed operators decided en masse to *buy* mobile operators instead of building a competitive infrastructure—they then lost any interest in building alternative networks. Meanwhile, the QoS and inter-Access-Point-roaming refinements to WiFi and the standardization of WiMAX took *much* longer than forecast, so the technology opportunity receded. As a result, the mobile operators looked over the abyss at the threat of voice over WiFi and WiMAX and decided it's not going to happen in the next planning cycle.

We can draw a helpful three-by-three matrix here. There are three main categories of player in the voice over wireless stakes: pure-play Internet telephony providers such as Skype, Vonage, and many others; traditional mobile operators

Table 2.2   Listing the Options—Type of Operator Against Type of VoIP Access

...	Skype-like VoIP Operator	Mobile Operator	Fixed Operator
Voice over WiFi/ WiMAX	Niche. Needs new handsets to enlarge market, but lacks ubiquity of coverage.	Tactical: UMA for cheap home-zone type service + better in-building coverage. IMS eventually.	Tactical: WiFi may replace DECT. A case for UMA?
VoIP over 3G data	Niche: needs HSDPA and HSUPA. Do economics stack up against circuit-switched mobile?	Eventually via IMS, once circuit-switches are removed post 2010.	N/A
VoIP over fixed access	Niche even with Vonage-like handset/ terminal adaptor. Hard to compete with CPS.	N/A	Niche until next-generation (IMS multimedia) network.

such as Vodafone, O2, Orange, and fixed operators such as BT, FT, DT. And then there are three possible ways to carry voice over IP—over a WiFi and/or WiMAX access network, over the 3G data Radio Access Network, or—as a baseline case—over a *fixed* IP access network. In the seven combinations of these, there are unfortunately no short-term voice-over-IP-over-wireless success stories (see Table 2.2).

If you are a mobile operator, then year-on-year things are looking not so bad. Your national RAN (radio access network) infrastructure and MSCs (mobile switching centers) move down through their depreciation cycles. The upside of all that expensive complexity is that mobile circuit-switched voice is actually carried rather efficiently and well. The fixed operators increasingly own or are buying back mobile operators, so are disinclined to invest in any kind of competitive infrastructure, as already mentioned.

There *are* people around who would like to undercut the cozy international mobile oligopoly, but the tools are just not there *now*. WiFi still needs a couple of years of work to get QoS (802.11e) and inter-AP roaming and authentication sorted out (802.11r). And wander away from the WiFi home or hot-spot and you're back in the arms of the mobile operator again. WiMAX 802.16e might make a difference, but it's a while away (come back in 2008 and we'll take another look).

Even when VoIP signaling successfully manages to traverse a WiFi or WiMAX network, it still needs to be managed by an end-to-end call-control layer. The only platform that can properly handle the integration of fixed (DSL, coax, fiber) + 3G data + WiFi/WiMAX access networks, along with the necessary service management and billing is IMS, and the releases that can do all that (release 7 and subsequent) will not enter service until 2009 at the earliest. Skype, Vonage, and other geek-friendly technologies may be chronically nipping at the ankles, but the mobile operators (think mostly carriers with both mobile and fixed divisions) believe that they have a free run through to 2009.

The threat on the horizon is that someone, e.g., from the Service Provider space, invests in metro, and core fiber + maybe some really fast DSL and festoons it with a ubiquitous access network of WiFi/WiMAX. Then they would add something like IMS-lite to make it all work end-to-end, and if the equipment prices were right, the cellular guys *could* be undercut. The encouraging deployment worldwide of metropolitan WiFi/WiMAX networks shows that this is beginning to occur.

The mobile operators are understandably not WiFi enthusiasts: it's a technology that doesn't fit at all into their 2G and 3G architectures or business models. However, as WiFi exists and has been paid for, it can—with effort perhaps—be exploited. UMA (Unlicensed Mobile Access), which will tunnel "proper, regular GSM voice/signaling" over WiFi, DSL broadband, and IP access networks has been developed by BT and a number of other operators and vendors (Figure 2.6). A special handset is required, which in addition to its normal cellular mode of operation can also support WiFi radio access.

When the dual-mode handset is in reach of a WiFi network, and is authenticated to log on to it, the handset accesses a UMA Network Controller (UNC) device via WiFi across the IP network. This UNC could be in the broadband supplier's network, or in the mobile operator's network. The UMA handset can embed its normal GSM voice and signaling traffic in a special protocol wrapper so it can be carried in IP packets over the WiFi network, and then the DSL

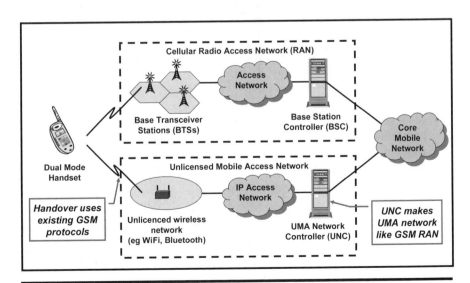

Figure 2.6   UMA architecture.

broadband network and then the IP access network through to the UNC. Here, the IP packets are terminated, the UMA protocol wrappers removed and the GSM traffic is conveyed to the mobile switch in TDM-mode (Time Division Multiplexing) exactly as if it had transited the cellular operator's normal Radio Access Network. Traffic from the network to the handset traverses the same route (in the opposite direction) using the same mechanisms. Normal GSM cell-handover mechanisms allow the handset to move seamlessly from the regular mobile network to and from the WiFi network without dropping calls.

Note that this is emphatically not "Voice over IP" where what is meant is that end-to-end voice is being carried as a native service by the IP network. This is conventional circuit-switched GSM voice that is merely being tunneled over the WiFi, DSL, and IP access network as it happens to be there. Ditto for the signaling, which remains ISDN-based. UMA is simply a way to emulate the RAN leg of the connection.

However, using UMA as a substitute RAN has a couple of advantages. First, it can be used to set special home tariffs in a precise way to attack fixed operators (where this is a commercial imperative). Second, it can be used to extend the mobile service indoors where the signal is often weak. UMA is therefore a tactical option that is on the mobile operator roadmap. An alternative is to plug a GSM pico cell into the DSL router, and boost the in-home GSM signal. The advantage is that any GSM handset will work with this solution, without having to be WiFi dual mode. The disadvantage is that the boxes are pricey, and it's not clear whether the customer is prepared to pay (Excuse me, isn't this pico cell thing part of the *operator's* national infrastructure?).

In conclusion, there will probably be a number of niche multi-access products:

- Dedicated WiFi handsets used in-house or in-enterprise,
- Dual-mode UMA-GSM/3G handsets,
- Early 3G dual circuit-switched voice + IP media/data IMS handsets

but there is no buzz about 2007, suggestive of something exponential in the offing. I think that realistically there are some major gating conditions that must be satisfied before the current paradigm tectonically shifts:

- We need to have a deployable IMS-type platform that permits roaming between DSL, fiber, cable, WiFi, WiMAX 802.16e, and 3G access networks. Something *cheap* like an open-source code-base would allow entrepreneurial companies to deploy it.
- We need the aforesaid entrepreneurial companies to *ubiquitously* deploy WiFi + WiMAX 802.16e using *cheap* equipment.

- We need handsets that can promiscuously access the various network types with smart-enough (IMS) clients.

Under these conditions, a genuinely mobile, ubiquitous Internet could emerge as a disruptive competitor to the mobile oligopoly, provided the regulators establish a level playing field, but don't expect miracles this side of 2010.

The Road Ahead

Now that the carrier vision of the Next-Generation Network is both in the global standards process, and has been committed to by all the major carriers, there is a tendency to believe the hype that it is indeed something new and innovatory, something different to the old-style telecoms networks, and indeed to the Internet itself. This is not *altogether* false. It is true that most of the NGN layers are simple assimilations of architectures and ideas that have been around for a long time. For example, the NGN simply leverages current thinking about IPv4/IPv6/MPLS networking and incorporates wholesale the .NET/JAVA EE architectures. IMS is genuinely new, however. There have been few substantial previous attempts to get SIP to work in a carrier environment, with all the necessary hooks to billing systems, authentication and authorization systems, plus links to a variety of service platforms and terminal devices. The IMS story is made more complex again by its genesis in the mobile cellular world, and the subsequent need for development to handle noncellular concerns such as wireless LAN/MAN, fixed broadband access networks and layer 3 mobility—currently work-in-progress as we have described. However, fixed-mobile convergence is part of the future reality we all have to manage and it is a huge advantage that IMS potentially provides a single standardized service architecture for both fixed and mobile operators.

Expect to see integrated NGN platforms successfully in service by 2008. For every technophilic early-adopter who will sign up with fleeter-footed start-ups racing ahead of the NGN, there are millions of consumers who couldn't care less about telecoms. They will be happy to wait for a consumer-friendly integrated service bundle from a branded incumbent carrier, with backup customer service in depth. The "get real" story is that the NGN will neither blow away the Internet, nor be killed by it. Three years out, anticipate a competitive landscape not unlike today, but with *significantly* better networks and services.

Appendix 1. The IMS Story

IMS has a reputation for complexity. Let me explain what IMS is and how it works. IMS stands for IP Multimedia Subsystem. IMS is the new signaling layer for next-generation networks that will allow services such as voice over IP, video-telephony, instant messaging, music downloads, "push-to-talk," and many others to be more easily brought to the market and billed.

IMS from All Angles

Brussels in Suburbia

You could mistake the hotel for an office block—it wears its three stars very lightly. The road outside is nondescript: all parked cars and delivery lorries. There is an excellent view of these from the restaurant.

I was in Belgium for a conference (technical telecoms conferences are almost invariably held in mid-range hotels), and, as the morning wore on, it occurred to me that one of the socially difficult things about conferences is lunch. Of course, it's all paid for, and the quality is invariably adequate, but, if you are attending the conference by yourself, when lunchtime arrives you will find yourself at a table full of strangers. This is called "networking."

Emerging from the morning session of the IMS conference, I decided, for no particular reason, to head straight for the restaurant, grabbing an empty table by the window. The view was largely blocked by a lorry parked outside. The first person to join me was Henry, who I recognized as a speaker from the recent session. Henry worked for a large American networking company and was active in the IETF. He was obviously Scott Adam's prototype for the UNIX Guru as he was tubby, bald, had a thick beard, and wore denim trousers with braces (suspenders, as American friends like to call them). His general demeanor? Well, you knew he was there. Henry sat down to my left, and immediately began digging around in his rucksack.

A rather diffident young man in a suit was the next to arrive, sitting down opposite me. Anton was from a Central European telco and worked in the research department. He was here to find out all about IMS for his company, and luckily for him spoke very good English.

My final companion was Sebastian, a handsome and energetic young man from a Mediterranean telco, who was expensively fitted out in business casual, and who casually explained to us that he worked in technical marketing, as he re-arranged his cutlery to my right.

Henry, Anton, and Sebastian. Are there never any women at these conferences? At a 95 percent confidence level, no. Actually I lie; the conference organizers are predominantly female, but as regards conference delegates, it's rare.

As we waited for the first course of the meal to arrive, Henry was intently working into something that looked like a WiFi-enabled PDA. I idly picked up some bread and asked Anton, the Central European researcher, what he thought of the conference so far.

"This IMS, it's very hard to understand," he began. "After the communists left in my country, we had Siemens, Ericsson, Alcatel, Nortel, and half a dozen other vendors telling us we had to change to digital switching. We spent millions and millions getting rid of our analogue and electro-mechanical switches—it was only a few years ago.

"Then we heard about the Internet and IP. The vendors came again and told us that the circuit switches had exploded. Now the call control was in a server called the 'soft switch' or 'media gateway controller,' the line cards were in devices called media gateways, and the switch itself—the matrix—had disappeared altogether and had become the IP network. We would have to buy a whole new set of equipment."

At this, Henry looked up from his plaything, eyes flickering at Anton and the rest of us, and gave a borderline feline grin.

Anton had been drawing on his napkin, and now showed us the diagram that marked the limits of his understanding (Figure 2.7), before continuing.

"Now the soft-switch seems to have vanished, and it's all IMS. I don't understand this IMS, what's it all about?" he complained.

The soup had arrived—something with vegetables and odd bits of meat, smelling vaguely of dishwater. My inner gourmet was easy to overcome as I waved the waiter away. My companions were less picky. As they worked their way through the soup, each according to his own national custom, I seized the chance to help Anton out.

"Don't worry," I said, "the soft-switch hasn't disappeared. It's just been relabeled as a 'call session control function,' abbreviated to CSCF. When people were originally talking about soft-switches a few years back, they were islands of IP in an ocean of traditional circuit switches and traditional phone handsets. But

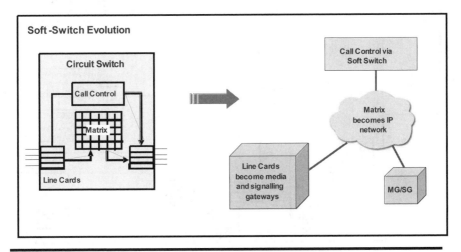

Figure 2.7   The evolution from circuit-switch to soft-switch

times move on, and now the emphasis is on smart phones, soft clients on PCs and so on. These devices send IP signaling directly to the soft-switch, and send their voice and video traffic directly to the IP network. In an all-IP world, the soft-switch story had to move on, and IMS is where it got to."

At this, Henry looked up and glared at me. "A typical ITU kludge." he mumbled under his breath, "Ridiculously overcomplicated, mind-numbingly complex." After this piece of invective his attention returned gloomily to his meal, while Anton, sharing his attention between me and the soup, nodded encouragingly.

"IMS is an architecture." I continued, "A way of splitting necessary functions between standardized components and defining very precisely how the components talk to each other. Let's starts with the CSCFs, the Call-Session Control Function components."

I began to draw my own diagram (Figure 2.8).

"When you use an IMS mobile phone, a PC or a PDA" (I risked a glance at Henry, but he seemed not to be listening) "it first of all contacts the local networks's Proxy CSCF (P-CSCF). It discovered this originally when it powered up and registered with the network, e.g. via DHCP. The P-CSCF is a special version of a soft-switch that is optimized to handle things like compressing signaling messages on the access link and encrypting them. The P-CSCF also polices operator policies such as disallowing certain soft codecs in your handset, and can instruct access routers to assign bandwidth for the call.

"The P-CSCF doesn't know anything about you as a subscriber, so it onward routes your signaling messages to your allocated Serving CSCF (S-CSCF) in your home network (this was also allocated at power-up and subsequent IMS

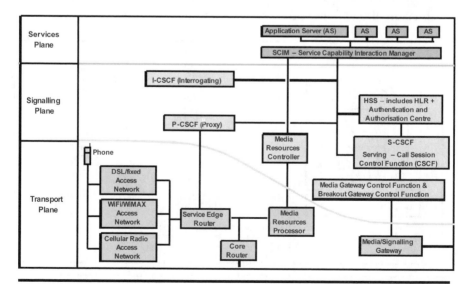

Figure 2.8   IMS architecture.

registration time). The S-CSCF has your service profile. The S-CSCF actually handles the call, bringing in application servers to provide you with value-added services (like the old Intelligent Network), finding the destination terminal, and managing charging.

"Now, Anton. Suppose you switched on an IMS phone here in Brussels. Your own network is, of course, out of reach back in your own country, so your phone would have to register via an IMS network here. Let's say it's Belgacom's network, so Belgacom would allocate you a P-CSCF. This P-CSCF would then act as an intermediary, facilitating the registration procedure between your IMS handset and the home S-CSCF of your operator back in your own country.

"Suppose you now call someone back home on your own network. Your signaling will first go to Belgacom's P-CSCF and will then be forwarded to your home-network S-CSCF—your handset stores your home network details. Your own home S-CSCF will forward your handset's signaling to the S-CSCF and then the P-CSCF, which look after the called-party. The called-party's P-CSCF then complete the call by ringing the called-party's IMS terminal. You're through.

"Suppose you were calling me in the UK, though, and I was a registered user of, say BT's network in the UK. Then your home-network S-CSCF would have to forward the call into BT's network to connect to me. To do that it would contact the third kind of CSCF, the Interrogating CSCF (I-CSCF) that guards the edge of operators' networks for third-party calls.

"The I-CSCF is like a border policeman—I'm sure you're familiar with those!—which shields BT's S-CSCFs from other operators' view. BT's I-CSCF will forward your signaling to the BT S-CSCF that knows about me. The specific BT S-CSCF that looks after me was assigned when I powered-up my phone and registered on the BT network. It got its information about me from the BT HSS (Home Subscriber Server, which stores my details permanently). My S-CSCF then forwards the signaling to my P-CSCF which will complete the call by ringing my IMS phone.

"The complete calling chain for signaling is this.

- Your IMS phone
- Belgacom P-CSCF
- Your network S-CSCF
- BT I-CSCF
- BT S-CSCF
- BT P-CSCF
- My IMS phone

Naturally, when we start talking, the voice over IP just goes from Belgacom to BT across normal routed links, quite separate from the path the signaling took."

Anton was nodding furiously, obviously following what was going on. Perhaps it was helping take his mind off the soup. Henry, however, looked at me with withering contempt.

"Almost all of that is wrong," he said. "The phone does NOT ring at that point, because there is a whole load of resource reservation to do first. The two terminals have a wonderful time exchanging still more messages to agree on codecs and bandwidth, and to reserve resources in the network, before anyone dares to alert the called party that maybe there is a call for them. The network drowns in messages. You know sometimes I doubt whether IMS will even work at all!"

"Henry, you know and I know that the specifications for setting up a call in IMS run to dozens of pages—there is no point drowning in detail when Anton here just wants to know how it works. Yes, there's plenty of messaging going on, and the messages themselves are pretty complex. But as you well know, there are a frighteningly large number of issues to be managed—multi-media capabilities of the terminals, user privacy, service customization, security, authentication, location...well, I could go on. The problem is complex, and therefore so is the solution. The argument is whether IMS is over-complex—I don't really think so, and if it is, then future releases will simplify it."

Henry looked like he was ready to argue the point, but with waiters hovering, he seemed to prefer to return to the last of his soup. Encouraged, I continued.

"Actually, Anton, you now know everything important about IMS. The other stuff is fill-in. The Media Gateway Control Function and Media and Signaling Gateways are the things you mentioned previously when you described the exploding circuit switch. They use IP technology to emulate a circuit-switch and connect to the existing circuit-switched telephone network, along with the Breakout Gateway Control Function, which controls where breakout to the PSTN should occur. You use these functions when calling someone on the existing phone network

"The Home Subscriber Server (HSS) is a database containing what the network knows about each user. For each signed-up customer, it holds authentication information, user service profiles, pointers to the billing function to use, and topical information such as where the user is, which S-CSCF is handling that user, and so on. The various CSCFs make extensive use of the HSS to authenticate users, manage services and route calls. CSCF identities are posted in the DNS, which is how one CSCF can find another.

"If you are going to make a three-way call, or your call gets network announcements like 'the line is busy,' this is the Media Resource Function operating. It can handle different media streams, providing conferencing and bridging, and transcoding, and can play announcements.

"Finally, the Application Servers (AS) are there to maintain any information which is needed to support advanced services. This could be anything from video clips, music files for download, network games servers, presence or location servers, and anything else you can think of.

"So you see, although IMS seems complex, it's mostly bringing together a bunch of functions which carriers need and then making sure it all works."

Anton glanced from me to my diagram, trying to put it all together as the soup bowls were removed.

"Of course, you're forgetting the most important thing," came an unexpected contribution from Sebastian on my right. He languidly made the letter "B" in the air with his finger, and smiled at me indulgently, as marketing people do to technologists.

"IMS is mostly about billing," he confided, "and most of the components just mentioned have hooks into the charging system. The CSCFs tell the charging system about what kind of call it was and how long it lasted, while the IMS-enabled routers tell the charging system about bandwidth used and how many bytes were transferred. It's all correlated to the same call, and ends up as a call detail record (CDR). As a result, for the first time we can do standardized online charging for pre-paid services, and offline charging for those with accounts with us, and even mix-n-match. Up to now we had to reinvent all this for every new service—you can imagine the expense!

"In fact, the Session Charging Function is our *favorite* example of an IMS Application Server."

Sebastian now adopted a mournful expression, addressing the whole table.

"The truth is, the telecoms industry is in a bad way. For years we lived off the voice revenues from our circuit-switched networks. Then the Internet came along, which we carried on our networks, but *we* never saw *any* benefit. Internet access was a cut-price commodity, and all the services were provided by other people at no apparent cost to the user. No wonder we're in trouble. With IMS thankfully we're back in the game, with services our customers will want to use, and that we can control and bill them for."

Sebastian sat back, looking rather pleased with himself, as the main course arrived. For a while, we ate in silence, although I couldn't help noticing Henry's rather aggressive use of the cutlery. Something was obviously bothering him. We soon found out what, for, as soon as the plates were cleared, Henry's angst became plain. It was with more than irritation that he glared at Sebastian and me.

"You guys are so typical of an industry that doesn't get it," he growled. "You were the guys who brought us decades of overpriced poor service, that backed ATM for the future, that fought against everything the Internet stood for."

I thought that was a bit unfair, I had been literally screamed at by a Nortel SVP back in 1998 for championing a customer's right to choose another vendor's routers for their new IP network, over Nortel's IP-enabled ATM switch. But Henry was relentless.

"Where were you guys when the Internet generated the most sustained period of innovation that the world has ever seen? Still does, at least for the time being. The Internet works because of its basic architectural principle, which you guys trample all over."

He looked at his PDA, did something with the stylus, and began to read aloud.

"Here, I quote: 'the (Internet) community believes that the goal is connectivity, the tool is the Internet Protocol, and the intelligence is end to end rather than hidden in the network.' That comes from RFC 1958 on Internet architecture from way back in 1996 (Carpenter 1996).

"It's because the network has the function only of transporting IP between end points that anyone with an idea has been empowered to innovate. We all know about Telco thinking: complex architectures taking years to produce that only they can use, fossilized and dead. The user has to take or leave it, and pay through the nose for the privilege. Innovation stops dead. You guys want to kill the Internet."

Then, with a final glare, "Someone should stop you!"

Henry seemed to be soliciting support from Anton, but for Sebastian and me, he reserved only scorn. I had heard this argument before, and a number of thoughts and rebuttals swam into my mind. But my tentative assembly of the components to an answer was cut short by a suave and practiced response from Sebastian.

"My dear Henry, you are muddling up so many issues. If you look at your Internet heroes—Google, Yahoo, Amazon—you will find that they are all using complex proprietary architectures with components and interfaces they do not make available to their customers. Their customers are forced to interact with them in a controlled way. The only difference between those guys and telcos with IMS is that IMS *is* standardized and *does* use open protocols—all, by the way, developed with the IETF.

"Now, just because a Service Provider is also an operator of IP networks—a telco—should they be forbidden from offering voice, video, and data services in a sophisticated way? Are you really saying that? Not very libertarian, is it?"

Anton was watching both parties, fascinated by the interplay of argument and personalities. Were these men going to start shouting or throwing things? The final course, a chocolate and cream delicacy arrived, and Henry struggled to eat and sustain his line of argument.

"Sophistry!" he fumed. "You guys will lock customers into a systems framework they will never get out of. Innovation will occur at your pace, which is to say—glacial, and you will charge your usual monopoly prices. The fat cats will continue to enjoy the cream, (at this, Sebastian raised his eyebrows at me) and meanwhile you will use your control of the network to suppress all competition!"

I had to admit that Henry had a point here. There had been disturbing accounts of new network appliances that could do stateful packet inspection on-the-fly and discard or impair low-cost VoIP suppliers such as Skype or Vonage, who did not use IMS. The new start-up Internet telephony players were unlikely to want to play settlement games with the big carrier IMS machines—and the latter's response might not be pretty—the equivalent of pizza wars, or sending the boys round to have a word with the competition. Sebastian, however, was having none of it.

"Libertarian scaremongering. It has always been possible to damage a competitor's traffic. Regulation and competition have always been the best defenses. IMS will make no difference. With IMS we will deliver a very sophisticated set of integrated services which scale and which work. Think of presence, voice, location-based services, video and music download, home-surveillance, networked games, and many other things all working on your fixed and mobile devices, wherever you are.

"It needs a powerful organization and a powerful architecture to deliver this kind of seamless quality service, and the telcos have it. Yes, you *will* have to pay, but competition will still exist—after all, everyone is doing IMS—and the business model of giving away services never worked in the longer-term for anyone.

"I hate to say this to an American, but—get real!"

As we drank our coffee, the discussion continued in a desultory way: the main points had all been made. I considered that it *could* all turn out for the best based on the following assumptions:

- that there was real competition for IMS-based services
- that market power and/or technology was not used to suppress the non-IMS competition, allowing new and potentially more flexible architectures to be deployed in parallel
- that the basic Internet, and unrestricted access to it, was not threatened.

You couldn't escape the fact that many of the services IMS promised do not exist today, not because the protocols and technologies aren't there, but because the services require a mammoth task of systems integration to get them to work and scale. IMS is the only game in town right now for that, but anyone else is free to come up with a rival architecture. So where is it?

Appendix 2. IP VPNs

This section is a lot more technical. Read it if you genuinely need to know the different ways an IP VPN can be set up and how the different versions work.

Today there are two main techniques that can be used to build an IP VPN: IP tunneling across the Internet, mostly using IPsec as the tunneling protocol, and BGP/MPLS VPNs—the true descendants of the Frame Relay and ATM VPNs, used by larger enterprises for many years.

IPsec VPNs

Coming from an ISP tradition, IPsec has traditionally been used to create encrypted, tunneled links across the public Internet between customer sites. The VPN topology can be full-mesh, or hub and spoke. IPsec, using a sufficiently-strong encryption method (e.g., 3DES, AES) can create effectively unbreakable encryption to suffice for adequate transmission security across the multiple Service Providers constituting the public Internet (Figure 2.9).

In an IPsec VPN, an IP packet from customer site A destined for a host at the remote customer site B is first routed to site A's IPsec Gateway. Here the packet and its header are encrypted, and the resulting data placed as a payload into another IP packet addressed to the IPsec gateway at site B. The latter packet is then routed to the public Internet where, just like every other IP packet, it hops from router to router along traffic-engineered paths until it is delivered to site B's security gateway. Here the "outer" packet terminates and its payload is decrypted to reproduce the original customer IP packet. This is then launched into the site B enterprise network for routing to its destination. The process is

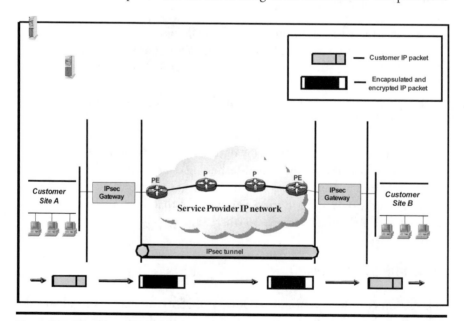

Figure 2.9   The IPsec VPN

exactly analogous to how company internal mail can be packaged up at the site A mail room for forwarding across the public postal system to be delivered to the remote site B destination mail room, where the package can be opened, and the individual office letters routed by the in-house mail system. To make the analogy more exact, we would have to assume one post office package per in-house envelope—no bundling!

A public Internet solution cannot guarantee QoS, where this is defined as relatively stringent bounds on delay, jitter, packet-loss, and path unavailability. However, carriers deploying on their own, adequately-engineered facilities could in principle use IPsec to implement VPNs with QoS guarantees. The major limitations were historically the inadequacies of the VPN end-point devices. These are the IPsec Security Gateways which are usually customer-premises or customer-located equipment. Since these emerged from an enterprise network past, it was difficult to find boxes with high-speed WAN interfaces and performance past OC-3/STM-1: however, the current generation of carrier-grade Security Gateways has much higher "wire-speed" encryption rates. To offer QoS, these devices would also need to copy customer QoS marking (e.g., Diffserv Code Points) from the host (customer) IP header into the encapsulating IP header for network processing. Note that Diffserv marking is usually managed by an enterprise router or QoS appliance, based on IP addresses, protocol Ids, port numbers, and perhaps deeper packet inspection. The host PC is rarely trusted to do QoS marking.

BGP/MPLS VPNs

IPsec VPNs have a "cheap and cheerful," almost DIY feel to them, that can suit the smaller enterprise. Larger enterprises seeking an IP VPN look to something which has more of the managed feel of Frame Relay or ATM VPNs. The solution is an alternative architecture for carrier-based IP VPNS, using Border Gateway Protocol (BGP) for managing customer routing information, and Multi-Protocol Label Switching (MPLS) to provide the private "tunnels" separating and forwarding each customer's traffic (Figure 2.10).

In this architecture, a packet originating from a computer at site A, and destined for site B is first routed to the Customer Edge (CE) router at site A. This can tell from the destination address that the packet needs to go somewhere nonlocal. The CE router has no responsibility for trying to decide which site that is, it simply forwards the packet to the Service Provider's Provider Edge (PE) router. The PE is really a collection of virtual routers, one per customer. The customer-specific virtual router has a routing and forwarding table which is specific to this particular customer, and knows to which remote PE router (and specifically which destination virtual router at that PE) the packet should go to. It attaches two MPLS labels to the packet: the inner one to identify that it is this

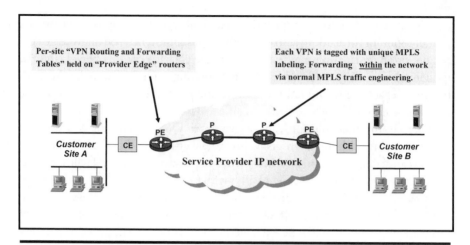

Figure 2.10. The BGP/MPLS VPN.

customer's VPN, and the outer one to specify a label-switched path through the Service Provider's network. The packet is then label-switched through the Service Provider network until it reaches the correct virtual router on the destination PE, which then sends the packet to the corresponding site B Customer Edge (CE) router, which, in its turn, forwards it on, within site B, to its destination. Note that the packet has at no time been encrypted: you trust the Service Provider to keep it safe, just as you had previously trusted it to managed its Frame Relay and ATM networks securely. The obvious question: how did the PE attached to site A's CE know where to send the incoming packet? How did it populate its routing and forwarding table?' The answer is discussed below. Since we have two quite different architectures for implementing IP VPN, the question is immediately raised as to the pros and cons of each.

The Role of the Edge Device

For BGP/MPLS VPNs, RFC 2547 defines the endpoint of the IP VPN to be a virtual router running on the Service Provider's "Provider Edge" physical router at the access PoP. This PE router hosts a number of virtual routers (more correctly called "Virtual Routing and Forwarding instances")—one for each customer-site connected. The customer-specific VPN is identified by the innermost MPLS label. Forwarding connectivity across the WAN (PE to PE) is provided via the standard traffic-engineered Label-Switched Path LSP within the Service Provider network, using the "outer label" for forwarding.

In the case of IPsec, the VPN end point is at the granularity of the individual customer site, and is physically a Security Gateway—typically a device incorporating a routing function with additional IPsec functionality (such as hardware encryption/decryption). The Security Gateways may be at the customer location—preferred for access security reasons, but costlier, or could be at the Service Provider PoP. In either case, the first Security Gateway encrypts customer VPN packets leaving the site and encapsulates them into "outer" IP packets addressed to the final Security Gateway. As the encrypted and encapsulated IP packet enters the Service Provider network and encounters the Service Provider ingress edge router, it is forwarded across the Service Provider network to the Service Provider egress edge-router along an MPLS traffic-engineered Label-Switched Path (LSP), en route to the egress Security Gateway. This assumes the Service Provider is using MPLS for traffic engineering, of course—most carriers are.

In the IPsec VPN solution, if the IPsec VPN was carried solely on the SP's own network infrastructure (*not* the public Internet), it would be possible to use the Authentication Header rather than the Encapsulating Security Payload option of IPsec, in tunnel mode, to carry the VPN traffic. This would obviate the expense and latency of encryption. Customer traffic would then rely upon the inherent security of the traffic-engineered MPLS Label-Switched Paths (LSPs) in the Provider network.

Routing Issues

Ignore discussion of VPNs for the moment and consider just the issue of how customers organize routing across their networks. Customers typically run an Interior Gateway Protocol (IGP) on their enterprise routers. Consider the case with a Frame Relay VPN, using Frame Relay point-to-point links in the WAN. Suppose we need to run our enterprise routing protocol over a Frame Relay WAN using Permanent Virtual Circuits (PVCs). We can do this in two modes:

- Non-Broadcast Multiple Access (NBMA). Here we have a full mesh of PVCs between the routers (and therefore customer sites).
- Point-to-Multipoint. For partial-meshes, we can treat the connections as serial links. Traffic then has to transit through multiple PVCs and routers to get across the WAN between sites where no direct connection exists.

A full mesh is expensive. it is also technically unfeasible when more then around 100 sites are involved, as there is too much overhead on the routers having to talk to so many router peers. Smaller sites, branch offices in particular, can only afford cheap, low-powered routers. So for big customers with hundreds or thousands of sites, a partial mesh has to be employed. This implies multiple hops across the WAN to get from source to at least some destinations. The points apply as much to IPsec as to Frame Relay and ATM. However, we pay a higher price in cost and latency terms in traversing IPsec tunnels than in traversing layer-2 PVCs as we have latency overhead associated with encryption, decryption, and encapsulation for tunneling. Clearly, this creates problems for remote-transactions and VoIP applications, where round-trip end-to-end delay budgets max out at 100 and 300 ms, respectively. For this reason, multiple IPsec hops are deprecated. Note that an IPsec VPN can have a hub-and-spoke topology without incurring multiple hops if the traffic pattern is always spoke to hub. The problem these days is that traffic is increasingly any to any.

The BGP/MPLS VPN architecture is characterized by a completely different approach to managing any-to-any site connectivity. Here, the CE router points a default route (0/0) to its connected port on the PE router. It's effectively pointing to the virtual router on the PE defined by the VPN Routing and Forwarding (VRF) instance assigned to that customer site. The PE in its turn receives site network information from its attached customer edge router via any convenient routing protocol (e.g., OSPF, BGP-4). The only objective of this peering is to allow the PE VRF table to acquire knowledge of the (aggregated) addresses at that site—it could even be done statically for a small site, by provisioning. As a separate exercise, each PE router exchanges the routes it knows about from its attached customer sites with the other PE routers with their own attached customer sites, segregated by VPN. This is done via IBGP route advertisements, using the BGP multiprotocol extensions to allow VPN tagging of routes to make them distinct. The "innermost" MPLS label—identifying the VPN—is also distributed by this means.

The bottom line is that the customer does not run a single routing domain over the whole of their network when using BGP/MPLS VPNs. Instead, each customer site is a unique and self-contained routing domain, with a default to the PE for off-site traffic. This allows massive VPNs to be built with thousands of sites. Internet traffic, as opposed to inter-site traffic, is directed by the PE to a firewall function on the basis of the customer IP destination address, and thence on to the public Internet.

BGP/MPLS VPN Is Too Limiting?

The BGP/MPLS VPN architecture only works within a well-defined MPLS domain. This has been taken to be a major problem for this architecture, but it isn't. It is debatable to what extent enterprises (such as multinational customers) will want their multinational VPNs hosted on more than one Service Provider anyway. However, even if they do, the architecture describes various ways of linking the VPN across Service Provider boundaries. As the BGP/MPLS VPN bandwagon begins to roll in earnest (and take-up has been remarkably good), increasingly Service Providers are agreeing to peer with each and interconnect their BGP/MPLS VPNs. There is also increasing interest in brokerage companies such as Nexagent and Vanco.

Another complaint is that BGP/MPLS doesn't work with dial-access. Again this is true. The solution is to interwork a tunneling technology (L2TP or IPsec) into the BGP/MPLS VPN: standard engineering methods apply. With most Multi-Service Providers lining up to support both BGP/MPLS and IPsec VPNs, interworking mechanisms will be required in any case.

The net result of all this work will arguably be to make deploying VPNs relatively simple. As an enterprise customer, all you have to do is to:

- Connect up your customer-site edge router (CE) to the nearest PE router.
- Get CE-PE peering to work (static, RIP, OSPF, BGP, for example, running on the above link).
- Get the Service Provider to datafill the PE with the VPN identification information.
- The VRF tables are then automatically populated: locally across the CE-PE link by whatever protocol is run on this link (or statically), and globally by IBGP-MP.
- The PE-P-P-PE label-switched path is set up automatically by a suitable MPLS label distribution protocol or mechanism.

This is, pretty much, autoprovisioning out of the box, leveraging the MPLS and BGP skills of the Service Provider.

At first sight, it looks like IPsec VPNs might need much more engineering. However, appearances could be deceptive. Each Security Gateway needs to be told which other Security Gateways are participating in the VPN via tunnel provisioning. This is a matter of suitable tools. Once it knows this, however, the CE routers/Security Gateways can auto-negotiate security associations using the Internet Key Exchange protocol, and once these are up, customer router discovery can then permit the propagation of customer network routing information across the VPN IPsec tunnels. Tunnel configuration is the hard part. The major issues are those which the VPN inherits from the scaling of the customer routing domain as discussed above, and any latency issues.

BGP/MPLS VPNs Trash the "Stupid Network" Internet Architecture?

I used to think this was a knock-down argument against the BGP/MPLS architecture. The IPsec VPN solution is manifestly well-behaved architecturally. It hosts the VPN on network-attached Security Gateways. As far as the Service Provider core network is concerned, it just sees IP packets (the outer encapsulating ones), which it treats just like any other IP traffic (namely forwards it along traffic-engineered LSPs).

However, in the BGP/MPLS VPN solution, the Provider Edge router actually maintains state about each customer's VPN routes in the VRF tables With private customer state information on the Service Provider network, this looks pretty bad for the "stupid network" model. However, things are not quite so black and white. After all, customer sites routinely advertise their network prefixes to Service Provider routers—that's how they become visible across the public Internet. So we already maintain customers' publicly-visible network state on the Service Provider network, some of which we later aggregate where we can. The BGP/MPLS VPN extra thing is to add—in a partitioned way—network prefix information which is NOT intended to be advertised within the public Internet, but only in scope of the VPN sites of that customer. Because such information is by definition customer-specific, this means multiple routing tables (the VRF tables). I wouldn't recommend dying in a ditch to preserve this distinction.

Is the BGP/MPLS Provider Edge Router Doing Too Much?

Well, it has a lot of information to store and maintain (for lots of customers) and we would need to watch the performance, size of VRF tables, upgrade strategies, single points of failure, and so forth. Also, if one customer suddenly brought a number of extra routes to the table (say, via an acquisition), it might hog router resources compromising other customers hosted on the same device. However, we need a sense of perspective. It would be a big customer site that advertised more than 500 separate prefixes. We might expect 30 or 40 distinct sites to be hosted off a particular PE: this gives a total of 20,000 routes max (assuming every one of them was very large).

However, the default-free Internet routing table is well in excess of 100,000 routes, and modern carrier routers take this in their stride assuming they were provisioned with sufficient memory. So not a show-stopper. It is worth pointing out that there are advantages for the Customer Edge (CE) router also to be running BGP. This may well stress the processing power and memory limitations of these smaller and cheaper devices, until the technology price-point catches up.

The IPsec VPN Security Gateways are also vulnerable, by the way. Either to size of mesh to be supported as the number of VPN sites increases, or to increases of line-rate, which stresses wire-speed en/decryption. Show-stopper? Probably not in the end.

Conclusions

A Service Provider interested in building serious, scalable, and QoS-conformant IP networks will find that BGP/MPLS VPNs are the way to go. As mentioned in the introduction, the BGP/MPLS solution is right out of the Frame Relay/ATM VPN stable, and has a similar "look and feel" to these earlier carrier-offerings, but with an IP spin. This sounds pretty good to conservative IT and network managers in the enterprise. By contrast, the IPsec solutions still have a marketplace brand of an "enterprise Internet solution," with all that connotes of small-scale, low-quality, down-market positioning and few, if any, service-quality guarantees. However, Service Providers of any size will have to offer both solutions. The BGP/MPLS solution plays well into the existing Frame Relay and ATM VPN markets, and can more easily handle large VPNs: those with many sites, and high data rates (> STM-1) on at least some of them. It also looks to be easier to manage and configure and is pushed hard by the router vendors, who cannot be ignored in market assessment.

The IPsec VPN solution is more of an "IP engineering" solution, but suffers from scaling problems and inadequacies of current implementations. It will suit customers who value high-grade encryption for their traffic, or who need a VPN which straddles multiple Service Providers. This applies particularly to extranets, e-commerce exchanges and companies with very mobile work forces. It can also be cheaper.

The explanations here are inevitably simplified. For more information on IP VPNs, see Pepelnjak and Guichard 2001–2003; Reddy 2005).

References

Bannister, J., Mather, P., and Coope, S. 2004. *Convergence technologies for 3G networks*, New York: Wiley.

Camarillo, G., and Garcia-Martin, M. A. 2004. *The 3G IP multimedia subsystem*, New York: Wiley.

Carpenter, B. (Ed.). 1996. Architectural principles of the internet, IETF. *RFC 1958*. http://www.ietf.org/rfc/rfc1958.txt.

Huitema, C. 2000. *Routing in the internet,*(2nd ed.). Upper Saddle River, NJ: Prentice Hall.

Pepelnjak, I., and Guichard, J. 2001–2003. *MPLS and VPN architectures*, Volumes 1 and 2. Cisco Press.

Reddy K. 2005. *Building MPLS-based broadband access VPNs*. Indianapolis, IN: Cisco Press.

Stern, T. E., Bala, K., and Ellinas, G. 1999. *Multiwavelength optical networks*. Upper Saddle River, NJ: Prentice Hall.

Chapter 3

The Next-Generation Network and TV

Much Ado about Broadcasting and Internet TV

October came around and I was at a client dinner in a traditional London club in Knightsbridge. Normally, I am not a huge fan of these events—they can go on late and it takes me almost two further hours to make the journey home. However, this evening was different: I was going to have an opportunity to meet Benedict, a leading television executive and long-time critic of the BBC. The dinner also promised the presence of a senior (and reputedly attractive) BBC strategy executive, Beatrice. I looked forward to both a learning opportunity, and a chance to watch the fireworks. Little did I anticipate that I was going to be the object of some entertainment myself, but more about that later.

As the dinner started, Beatrice, sitting diagonally opposite to my right, was the center of attention—the only woman amongst twelve suited and middle-aged men. Benedict, sitting to my left, had been involved with a policy working group associated with the UK Conservative Party that had been tasked to take a look at the BBC and come up with some thoughts about its future. I had read the group's report (Broadcasting Policy Group 2006) and, introducing myself to Benedict, asked him to explain what it was all about.

Benedict was only too pleased to oblige, and, as we intensely engaged with each other, I was already committing a cardinal crime at a client event—monopolizing one guest and ignoring everything else going on. Oblivious to the increasingly irritated glances I was getting from my colleagues, I hunkered down and listened to what Benedict had to say.

The BBC, Benedict explained, had not always been the lofty establishment pillar it is today. It was founded in 1922 by wireless manufacturers to make radio programs and so encourage the sale of their products. Five years later it was nationalized under its first general manager, John Reith, as the government began to understand the implications of broadcasting both for national security and also for the "cultural improvement of the masses."

As the radio and later TV industries developed, the BBC had always fought against competition: it sought to retain its monopoly of broadcasting in the Beveridge Inquiry of 1949, it opposed the launch of the commercial Channel 4, and campaigned against cable. Its funding by the license fee (currently just under £130 [$250] per year) is a flat tax that bears most heavily on the poorest, and is levied per-TV-owning household regardless of how much BBC programming is actually watched.

It ought to be easy to dislike the BBC, Benedict said. It is a large, publicly-owned, bureaucratic, vertically-integrated, slow-moving monolith with significant market power. Its very existence suppresses the independent production sector and distorts the market. Yet somehow, the BBC is widely admired and respected across the world. Even avid free marketeers mutter that although the BBC in its current form would never be invented today, as it's here, it would be a mistake to abolish it. Hearing this, my free-market impulses could be restrained no longer.

"Benedict," I declared, "there is nothing the BBC does that isn't being done equally-well on commercial channels, in this country and abroad. It's ridiculous that we're taxed through the license fee. The BBC should be abolished forthwith and people should be free to choose whichever programs they want through normal market mechanisms!"

Benedict smiled at me, with mocking pleasure. "What about programs that are merit goods?"

I reflected on this for a moment. The term "'merit good'" is economists' jargon for a good that people allegedly undervalue because they narrowly only see the benefit to themselves, not the additional benefits that their consumption generates for others. Examples of merit goods in broadcasting include news, current affairs, politics, history, science and high-art. Except for people disparaged as "intellectuals" in Anglo-Saxon countries, it is widely felt that most citizens are less interested in these topics than they ought to be in support of an informed democracy. Few people make a similar argument for soap operas, quiz shows, and sports (except for cricket in the UK). But then, what is so special about TV, I asked myself?

"I don't see your problem." I replied. "You can go down to any news stand and buy quality publications like the *Financial Times* or *The Economist*. You can also buy sports papers and top shelf magazines. It's entirely up to you, no one is forcing you to pay a newspaper license fee to subsidize politics or 'culture'."

Benedict thought for a moment.

"Yes, a good libertarian reply, but you're missing some subtleties. A person may choose to eat junk food, listen to junk music, and read nothing but junk, but most people would accept that they're making choices that physically, culturally, and intellectually impoverish them. But in the case of TV, the person paying for the channels is not necessarily the only one watching: there are wives, husbands, and children who also watch and listen. TV is immediate and pervasive. Sparks of news, science and art can catch fire and change lives, despite prior ignorance or lack of interest.

"TV and radio are also *experience goods*. It's difficult to assess the value in advance. Many people would reject new cultural experiences if they had to pay upfront. By making the marginal cost of such programs zero, we encourage them to take a look, and sometimes they surprise themselves by liking what they see. We owe it to each other to sow such seeds, even if many of them never take root."

I was still skeptical that in this culture-soaked world of ours that such heroic efforts were really necessary—just look at the magazines in any newsagent, the DVDs in any megastore and the infinite riches on the web, but I could see Benedict would not be convinced. I tried a different tack.

"You are not a fan of the BBC, but you accept the merit good argument. How would you get merit goods made by commercial broadcasters without endless regulation?"

"Usually merit goods are supported by government subsidy. This increases supply to the 'socially necessary amount'. Other countries have set up the equivalent of 'Public Service Broadcasting Boards' that dispense public funds. Program makers and broadcasters pitch concepts to these in the search for funding. Since some of the cost of the program is subsidized, the industry has an incentive to add such programs to their portfolio mix. High-production values are also catered for, because production quality is a key differentiator in a competitive TV and radio market anyway."

I thought this was a good point—if you *have* a concept of merit goods, then there was no reason to believe that this approach wouldn't get them made. I considered some of the implications while looking around. By some magic we had managed to eat the hors d'oeuvres and the main course without really looking at them. On my right was an American executive who I had totally ignored up to this point. He had picked up on fragments of our conversation and was not keen on this BBC bashing.

"Where I come from, TV is garbage." he suggested. "You need to think real carefully before you do anything to damage the BBC!"

The dross argument . . . I turned it back to Benedict. "He has a point. U.S. TV is normally cited as the existence proof that pure commercial TV is a race to the bottom in terms of quality. The programs are terrible, and are almost unwatchable due to the frequency and length of the ad breaks." I had spent two years in Vienna, Virginia, and so I knew what I was talking about. Benedict paused, as if

I had delivered too many confusions in one breath to easily deal with. He began to tick off points with his fingers.

"One, there *are* ad-free subscription channels in the States—Home Box Office comes to mind—which produce programs that are generally considered the equal or superior to any program produced on UK TV. There *is* an audience for quality of content, and subscription is a way to fund it.

"Two, people say that free-to-air channels funded by advertisement is an inefficient form of advertising because TV is totally one-way and undiscriminating. Because the ads are untargeted, the largest possible audience is sought, they say, and because the average viewer is likely to be not that interested or engaged, ad rates per thousand viewers are low. This tempts the broadcaster to schedule lowest-common-denominator programming to scoop up the greatest possible audience size. Well, that's the theory, but it's not a very good theory. Consider that newspapers are largely funded by advertisers, but manage to differentiate in terms of 'quality' quite successfully. And commercial channels in the UK, lightly-regulated for ad-frequency, merit-good content and scheduling, do well against the model of perfection offered by the BBC."

Benedict hadn't quite filled in the bottom line, so I figured I should do it for him. I turned to my American friend and attempted to summarize.

"What Benedict is saying is that first, there *is* a market for higher quality material quite independent of whether the BBC is around to mandate it. HBO and similar channels show that. Second, commercial models, sweetened by some form of merit-good funding and light regulation on advertising *can* deliver a quality viewing experience every bit as good as the BBC. The so-called unique merits of the BBC are not so unique after all. And Benedict is quite vocal on the demerits of a BBC-like organization in terms of locking up talent, distorting the market and limiting creativity and choice, quite independent of the iniquities of the license fee itself."

Benedict seemed content with my summing up, and the American was distracted as we had by now finished our meal and a more general discussion was opening up.

Beatrice, the BBC strategist, was outlining the BBC's plans for the future. All bases were to be covered: free-to-view digital channels as well as the BBC's Internet platform and future video-on-demand. Only an organization of the size and capability of the BBC could hope to propel British Broadcasting to this modernized future.

Excuse me? Why would we want to rely upon a monolithic, bureaucratic monopoly to pioneer Internet TV? Benedict was making some tentative demurral but in my disdain, I overrode him. I addressed myself to Beatrice with barely-concealed scorn.

"Excuse me, Beatrice. Benedict may be inhibited about criticizing the BBC, but I have no such reservations. The Internet is going to completely disintermediate you—you have no chance of riding that particular tiger!

"Any product company can create streaming or downloadable TV content, and DRM is good enough to do effective rights management. The costs of entry into the portal market are low—we can expect many Internet portals offering video-on-demand, analogous to today's channels,. The BBC *may* be a major player in this future, through inertia, but it is surely not necessary!"

I stopped, and wondered at myself. I was excitable and sweating, my heart thumping. This had been a rant, not an urbane after-dinner conversation with a client.

Beatrice turned to me, smiling sweetly. "Thank you for your views. We *have* thought a lot about these issues, and we are sure the BBC can continue to add a great deal of value to viewers in these areas as we have in the past." Then her glance moved on, as she continued talking to the other guests: it was like a torch beam had been switched off.

The magnitude of my error began to hit home. I had behaved like a gauche heckler at a public meeting rather than a consultant at a client event. I had attacked an important client in a most intemperate way in public. I looked for a hole to climb into and failed to find it.

I e-mailed Benedict the next day, hoping he could be my intermediary in communicating an apology to Beatrice. He reassured me "She's a tough operator, used to dealing with criticism. I doubt she remembers, anyway." That last sentence pretty much defined the evening for me.

Experience = Bits Per Second

Let me run past you a fairly ambitious statement: any human experience can be delivered via a bit stream. This truth ought to open up a realm of possibilities to carriers capable of delivering such bit streams. Experiences such as touch, taste, and scent are not included in most carriers' product catalogues today because we neither know the encoding rules, nor do we have the right interface technologies. These are scientific and engineering problems that will not, however, remain unsolved for ever. They are, for example, research goals in Japan's "3D TV" project. Speech, by contrast, has long been a staple of the carriers' portfolios, in the shape of the PSTN, and latterly in an improved form with IP telephony, with its greater bandwidth.

The hottest new area is video. Historically the bandwidth required to deliver an acceptable video service had been beyond the abilities of the carriers' access networks. With the arrival of DSL broadband technologies, this changed and it has become possible to transmit even high definition video to a large fraction of a carrier's customers who are not too far from an exchange (signal attenuation causes bandwidth to drop as the copper loop length increases). Carriers with access to fiber-to-the-home, or hybrid fiber-coax/copper to the home, can deliver even higher bandwidths, although these more modern access networks are extremely expensive to build-out.

Just because you can technically carry video to customers on your new broadband networks doesn't mean you have a business case. TV has been around for a while, and has a complex and mature value chain all of its own. There is a limit to how destabilizing the Internet will prove to be, and the existing players have proven themselves to be formidably adaptive. We will therefore next briefly review the TV industry, and then look at some of the lessons of Internet distribution of consumer products. Next we will consider some of the new services which can exploit the two-way capabilities of Internet broadband access, services which are unavailable in a pure one-to-many broadcast model. Finally, we will examine the mechanisms of TV over the Internet, and try to ascertain where carriers can profitably play. Business strategies for the major players are discussed in the final chapters of this book.

The Traditional TV Value Chain

Carriers contemplating entering the TV over broadband market have some thinking to do about what their differentiators might be. TV in its broadcast variety is a mature service, and is already delivered over terrestrial, satellite, and cable networks. Moreover, each of these transmission modes are highly efficient for one-to-many distribution, and can carry much greater bandwidth than DSL, allowing hundreds of simultaneous channels to be delivered. While no one can watch that many channels at the same time, there are often many consumers of TV in the same house (including video recorders) and people value the ability to rapidly flip through the channels on offer.

It gets worse. The TV industry is highly structured. The value chain can be analyzed in various ways, one of which is shown in Figure 3.1.

- Content Creation is often done to order by production companies.
- The Content Owner is the holder of the original rights to it: it could be a studio, it could be a governing body (e.g., a football association).
- The Content Aggregator is an entity which selects, acquires and edits content. It could be a channel (e.g., the Discovery Channel), an existing broadcaster (the BBC, Sky), or a Web portal.
- The Service Provider is the entity that owns the customer relationship and bills for services: typically a broadcasting company (Sky, BBC).

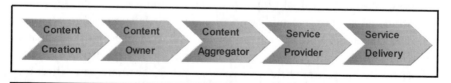

Figure 3.1   The traditional media content value chain.

■ Finally Service Delivery is accomplished by an organization running a suitable platform, e.g., Astra for satellite transmission in Europe, Crown Castle for BBC's digital transmission. The carriers now feel they can enter this space with their new broadband networks.

Many vertically-integrated organizations like the UK's BBC have historically internalized the complete value chain. The tendency today, however, is towards disaggregation and separation.

Why Do Channels Exist?

A broadcast medium, such as satellite direct-to-home, traditional terrestrial radio, or cable company coax, offers the opportunity for multiple parallel TV programming (each transmitted program uses only a fraction of the available bandwidth). There is, however, no way to tailor fine-grained content directly to individual consumer requirements. Instead, aggregate consumer demand is partitioned into a number of different ensemble-products targeted at different parts of the market, with distinct revenue and cost profiles. These products are called channels.

Because demand for different channels is often inversely-correlated (see below), it makes sense for Service Providers to sell channels in bundles rather than individually. Because of the large fixed costs involved in being a TV Service Provider, and spectrum scarcity, competition is limited and customers have to select among the bundles on offer.

Why Are Channels Bundled?

Suppose there are two channels, Sports and Arts. Stereotypically, Angela values the sports channel at $50 per annum, but would pay $100 per annum for the arts channel. Bruce has the contrary valuation (Figure 3.2).

■ If we price both channels at $50, each party will buy both channels for a total revenue of $200.
■ If we price both channels at $100, each party will buy just their most preferred channel, for a total revenue again of $200.
■ However, if we bundle both channels together for $150, each party will buy the combined bundle, and this time revenues are $300.

In this example, channel bundling has reduced the variance of willingness to pay, thereby increasing revenues by 50 percent for no extra cost. In this case, both parties get what they want at optimal costs to themselves. However, the theory of bundling throws up many cases where the customer is forced to buy the bundle and

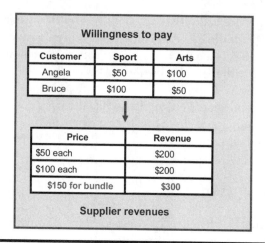

Figure 3.2   Bundling channels.

is unable to buy just the specific products they wish. In particular, a dross channel devoted purely to advertising, fund-raising, or shopping can be forcibly bundled with more interesting channels. This tactic can improve profits for a supplier with some market power at the expense of further customer dissatisfaction.

Channels as a Lowest-Common-Denominator

Channels are *averaged,* even *lowest-common-denominator* products, and this detracts from the value consumers place upon them. This limits the amount customers are prepared to pay for a channel, even before bundling reduces the value still further. Subscription-based charging is thus inhibited, so in many cases a free-to-air model is adopted, funded by advertising.

Because the advertisements are themselves not targeted, they are often perceived by end-users as intrusive and irritating, and this further lowers the value of the channel. Advertisers, recognizing this trend, tend to pay low, bulk rates on the crude metric of number of viewers (cost per mille = cost per thousand "impressions" or views).

Lessons from Internet Retailing

Given this level of customer dissatisfaction with linear broadcast television, there is arguably a real opportunity for the kind of mass-customization that Video-on-Demand, TV over the Internet, could provide. How would this work?

The Internet has so far developed two solutions for non-TV content aggregation and distribution: the portal and the search engine. The two are not at all

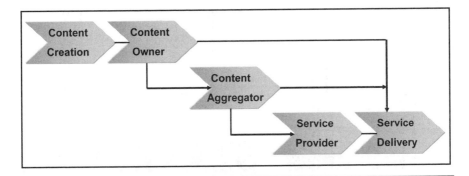

Figure 3.3. Video-on-Demand Value Net.

counterposed: a portal site that maintains a structured catalogue will generally also provide a search engine. Amazon is a classic portal: it organizes its collection of books-in-print into a comprehensive structured catalogue, and also provides the ability to search by keyword. Retailers of software and music, products which lend themselves to Internet distribution, have adopted similar models.

The ease of Internet publication seems to open up options within the value chain we saw earlier. It transforms, through disintermediation, into a value net (Figure 3.3), where certain upstream players can go "straight to publication." We are already seeing similar phenomena for book self-publishing, with book printing on-demand, and bands publishing their music directly to the net, or to low-cost portals specializing in breaking new acts. This disintermediation has its limits, however. The mainstream audience has expectations of quality content and high production values, attributes guaranteed by the mainstream broadcasters with their strong resource base and brand identities. It still needs major marketing muscle to bring even very good content to a mass audience.

Because of the vertical structure of the TV industry, the major broadcasters (Sky, BBC, ITV) own rights to extensive archives of content. DVDs have already provided them with a new channel to market to extract further revenues from these back-catalogues. Internet VoD will provide them with another. Expect agreements between these companies and major carriers to set up branded portals, or mergers between broadcasters and facilities-based carriers in another form of convergence. Once they get established, there is room for new content development models—pilot shows can be launched via VoD, which may well provide broadcasters with much better information about audience responses and demographics than they get at the moment.

The "Long Tail" of Saleable Content

In October 2004 Chris Anderson published an article called "The Long Tail" in *Wired* magazine. Anderson was calling attention to the phenomenon whereby

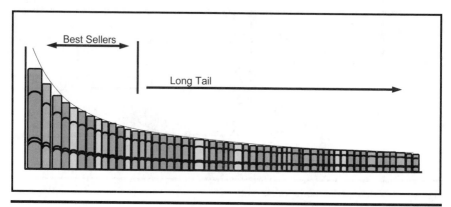

Figure 3.4 The Long Tail.

Internet stores are not bound by inventory restrictions. Physical stores, by contrast, *do* have to worry about shelf-space costs and per-square-foot revenues.

As *The Economist* (May 5, 2005) observed, commenting on Anderson's thesis in 2005, "In the case of Amazon, for example, around a third of its sales come from outside its top 130,000 titles. Similarly, Rhapsody, a streaming-music service, streams more tracks outside its top 10,000 tunes than inside" (Figure 3.4).

The focus of publishing companies—print and media—on blockbusters is revealed as partially an artifact of the traditional cost structure, which rewards a few big selling items far more than a large number of modest selling items, even though the overall revenues may be much the same.

Making a business from the long tail of modestly-selling titles requires some new thinking. Both Anderson's original article, and *The Economist's* review identify the difficulties of customers finding content in a large catalogue. By hypothesis, there is a large amount of content any particular customer would appreciate and purchase, if they only knew it existed and could access it.

Personalized content filters are one answer. Amazon's self-service technique is well-known—customers are presented with recommendations based on other, similar, customer buying histories: "Customers who bought this item also bought. . . ." Undoubtedly, there is scope for further innovation here—the mathematics of compiling such recommendations is similar in complexity to that of the page-ranking algorithms used by Google, and the industry has a way to go.

Another challenge is pricing. Amazon sells books and similar physical objects that are priced individually. However, sellers of music tracks and videos are selling digital information with low marginal *cost to provide* (most of the costs are whole-sale charges from the rights owners, estimated by Anderson at around 65 cents per track). Given a basic inventory of value to many people (the blockbusters)—plus the "long tail" of items where an item's value to any particular customer may be quite unpredictable, how should access to this inventory be priced?

It is possible to adopt a standard pricing, such as 99 cents per track. However, another, perhaps more sophisticated model, is to offer a two-part tariff, with a fixed price (e.g., $14.95 per month) for admission to the archive, and then a smaller, or zero, per-track price. Here is a model for how a customer might value downloadable songs (the same argument would apply to TV programming such as videos and films, games, etc.). There are a number of songs the customer really likes, and would be prepared to pay a lot for. Successive songs are less preferred and the customer would be prepared to pay less, down to songs where the few pennies of possible price are outweighed by the costs in time and effort to download in the first place.

If all the customers agreed on the same rank order of tunes, and had equality of income and desire, then the company could price each song at the market rate and make maximum profits—perfect price discrimination. However, customers vary widely both in their preferences and their willingness to pay. Everyone has their own idiosyncratic per-track valuation curve.

In this situation, it can pay to introduce a fixed tariff, a subscription. The subscription covers the value each customer ascribes to his or her personally-highly-rated tracks, so the customer is prepared to pay it. The cost per track can now be set very low to encourage the customer to continue to purchase, and to encourage lock-in. If the marginal cost is set to zero, then the fixed tariff can purchase an "all-you-can-eat" service of unlimited downloads, restrained only by the cost each customer puts on their own time and effort. All of the major legal downloading sites are adopting this model, in addition to selling tracks individually. Digital Rights Management is needed to prevent customer arbitrage (i.e., onwards publication of the material without the fixed tariff fee), but that is a story for later (chapter 11).

In summary, use of the Internet as a retailing channel has transformed the back-catalogue into a potent revenue stream. The key to unlocking it is (1) powerful search, clustering, and recommending systems; (2) innovative pricing schemes to encourage incremental sales; (3) a usable DRM system that protects content rights.

The Possibilities of Two-Way Services

We mentioned earlier that broadband offered a high-speed return channel, unlike existing broadcasting platforms. So far, the only use we have found for this is returning superior feedback to broadcasters and advertisers, and allowing programs to be ordered on demand.

Many commentators have argued that broadband multimedia will catalyze a new kind of entertainment, called interactive multimedia. It sometimes surprises people to be told that this is not some terra incognita glimpsed only dimly through the mists of future time: interactive multimedia is already here, and we call it networked gaming.

Gaming has a poor reputation, based on stereotypes of first-person shooters like Quake, paeans to gangster culture like GTA and immersive alternatives to a life such as Everquest, World of Warcraft, and Second Life. The stereotypical user is a male in his teens or twenties, addicted to surrogate violence and with too much time on his hands. And, by the way, the level of audio-visual quality in these games is getting truly stunning.

Gaming is perceived to be niche because the stereotypes have a great deal of truth behind them. On the other hand, there are nurturing games like The Sims, and quest games like Myst and Riven which seem to appeal to a much wider audience (this is perhaps a euphemism for the fact that girls like to play them as well). Simulation games too, such as flight simulators, or historical re-enactments, are not always assimilable to the "mindless violence" strand of the gaming market. An emerging niche is that of "casual gamers" playing arcade-like games (Solitaire, Bejewelled) (International Game Developers Association 2005). These are usually network-hosted games, rather than downloads or retailed CDs. And the demographic is interesting: the players appear to be mostly elderly women.

The gaming industry is unsure of the future (Rollings and Morris 2005). Market expansion away from hard-core gamers seems to require games that you can dip in and out of in episodes, rather than requiring a huge investment of continuous time. Existing game architectures do not lend themselves to this usage model.

A final point on games—the audience is not only consumer, the military has a long history of using simulations, see Wray et al., 2005, where virtual-reality environments, often distributed, link troops to simulated opponents. These often use the latest in AI-based cognitive modeling technologies. Oligopolistic markets, where competitors are few and large, and where outcomes are dependent on the actions of known competitors would seem to lend themselves to analogous simulations, given relatively modest improvements in technology.

Multimedia "New Wave" Products

The most obvious product that exploits the symmetric bandwidth of broadband and multimedia is plain old video-telephony. It has been a truism in the business that no one wants this product. Repeated attempts to introduce video-telephones have failed. People apparently want to talk, but not to see and be seen.

I wonder whether the problem is more that a threshold of usability has not yet been reached? It's arguable that if video calls were easy to set up and the camera and screen generated an experience of standard color TV quality, the service might take off. At the moment, we lack session management systems, such as IMS, sufficient bandwidth—particularly upstream and affordable terminals to make this easy for the average person, so the jury is still out.

There are a number of other potential services that await sufficient bandwidth and the development and integration of the appropriate terminal technologies.

- Flat screen virtual windows fed from a remote camera, perhaps showing a tropical beach scene, or a mountain view. There has to be a rental service here waiting to take off.
- Video wallpaper allowing a remote location to appear to replace a wall—a kind of virtual room extension. This might make videoconferencing more of a replacement to traveling, and for most scenes compression should tame the potential for bandwidth explosion.
- Genuinely immersive virtual reality. Like power generation through nuclear fusion, this always seems to be about to happen. Like many much-anticipated innovations, immersive virtual reality is the integration of many difficult technologies: 4D object modeling and rendering; user position and motion tracking; terminal devices interfacing to eyes, ears, etc; lightweight, low-power, and tetherless equipment; enough speed to do all of the above, and an affordable price. I guess it is no wonder we're not quite there yet, but when we are, as a platform technology akin to the invention of the laser, it will transform everything.

As is customary in a "family book," I pass over the extent to which these markets will, in fact, be driven by "adult content."

IPTV and VoD—Making It Happen

The next-generation network, with its IP transport protocols and broadband access has made it technically possible to carry TV programming, thus opening up a new business opportunity for carriers and Internet service providers. But what exactly is the product? In today's world, totally dominated by broadcast TV models, it's usually considered that there are three services.

First, we have IPTV. This means the service of offering a number of linear TV channels over an IP infrastructure—essentially identical to that delivered over other platforms such as satellite, cable, or terrestrial transmission.

Second is Video-on-Demand (VoD). This means the creation of a structured set of material: typically films, light entertainment, documentaries, adult material, and the like that is stored on video servers and can be accessed on-demand. This could be a free service, or the customer could buy a subscription package, or individual titles could be purchased on impulse.

Third is "Catch-Up TV." CUTV, is a form of VoD that archives say, the last week's programming from the linear IPTV channels. If you missed your favorite Soap or documentary, then you can pull it down later from the CUTV service. It is distinguished from straightforward VoD for several reasons:

■ Its evanescence—material is transient and may drop out after a period.
■ Its sheer volume—there is a lot of programming to capture and store.
■ Rights issues—the broadcaster may not own rights to store all broadcast material.
■ Content management—it may be hard to access broadcast material from third parties in a form acceptable for caching: there may be issues of quality, security, and metadata availability.
■ User interface—providing an EPG (electronic program guide) enhancement to navigate around so much material may be difficult.

What does a carrier have to do to get into the TV distribution business and provide these three services? Like the VoIP discussed in the previous chapter, TV is an overlay network that exploits the underlying IP network. Each of the three services mentioned above exploits the same basic architecture. There is a head end that assembles and prepares the media. This is then played out across the IP network. Finally the TV over IP media streams are received in the home over the broadband link and reassembled into programs on the TV, or perhaps a PC. A PC can do this itself, but the TV usually needs a special Set-Top Box (STB) as decoder. We will look at the process in a little more detail with reference to Figure 3.5.

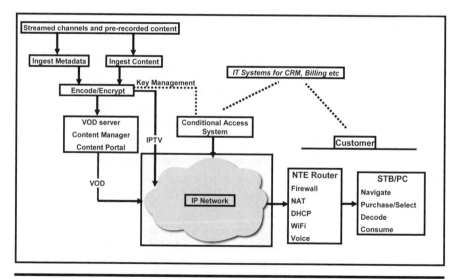

Figure 3.5 IPTV/VOD architecture.

At the start of the transmission chain, media will be delivered to the carrier in a variety of different ways and formats. Sometimes the media is on videotape, sometimes already on video servers, and sometimes it can be received "off-air" from a satellite dish or communications link. The process of acquiring content and its metadata is called "ingestion."

An ingest system will be able to process videotape, video material from servers and material taken off-air. It will be able to handle multiple data formats and transcode them to the formats required for further editing or transmission. It will be able to take its input under direct user control, or via a pre-loaded schedule, or as batch processing. Metadata as well as content is ingested, and this supports program browsing, verification, and editing.

Broadcasting involves *scheduling*: not just of programs but also of advertising and interstitials (promos, channel, and sponsor idents, etc.). Scheduling systems are often complex pieces of automation, and directly drive playout systems. Playout is the process whereby material, under scheduling control, is taken from an ingested source (e.g., a video server, or in real-time off-air) and is then supplied to the transmission system. For an IP network, the ingest system should already have transcoded the program material into a suitable (compressed) format (e.g., MPEG-2, MPEG-4) and this now needs to be encapsulated into IP and streamed onto the network. This is done by suitable hardware equipped with the right kinds of line cards. Once received in the home, the MPEG/IP stream is processed by the STB or PC to reacquire the TV signal, and this is shown on the screen.

Conditional Access Systems

The above describes the simplest case: free-to-air IPTV. For pay-TV, the programming is encrypted, and can only be accessed by the customer once a fee has been paid. Content encryption and decryption (often called scrambling and descrambling) is straightforward and is carried out by modules at the head end, and within the PC/STB. The harder part is key management, which forms the heart of the conditional access system.

The encryption/decryption keys, called "control words," are used to encrypt and decrypt the TV data stream. A control word is changed regularly every 10–30 seconds. The control words, in an encrypted form, are sent at sub-second intervals to the user within a parallel MPEG message channel as Entitlement Control Messages (ECMs). The high repetition rate is to ensure a rapid decode once a channel is selected by the user. These ECMs are received by the conditional access module in the STB, comprised of some combination of STB specialized hardware and a smart-card inserted into the STB.

Recall that the control words are themselves encrypted—it would make no sense to sent decoding keys in clear. The encrypted control words are therefore decoded within the STB by a service key which is centrally distributed to the smart-card perhaps monthly (again in an encrypted form: the recursion ends by the smart-card having a hard-wired decode key for this purpose). Since the STB can first of all decode control words and then use them to decode encrypted programming, what is to prevent the customer from viewing everything that is encrypted, whether they have paid for it or not?

The answer is that the STB won't let them. A customer has first to purchase entitlements to decode content. When the customer has paid for a service via the billing system, they are issued with a specific authorization to view what they have paid for (usually by the head-end subscription management system). This authorization is delivered to the STB in the form of Entitlement Management Messages (EMMs), which are also conveyed within the MPEG transport stream—note that although every STB sees every EMM, it can pick out those that are specific to itself. And it won't decode without one. For a more detailed treatment of conditional access systems see Tranter 2004.

It is clear that the conditional access module in the STB is a powerful gatekeeper. Once a broadcaster has persuaded customers to invest in their STBs, other broadcasters could be locked out unless customers are prepared to buy and attach multiple STBs to their TVs. To improve competitiveness, the Digital Video Broadcasting (DVB) standard group developed the Simulcrypt standard, which defines an architecture for conditional access systems that standardizes both scrambling algorithms and control word management. Conditional access head-ends and STBs built in accordance with the standard can be used to receive programming from multiple broadcasters, or alternatively can permit the broadcaster to upgrade their CA system in a modular fashion.

Conditional Access and Video-on-Demand

Video-on-Demand requires in the first place that archival material should be stored. This is achieved by placing content on scalable video servers, often in a pre-encrypted form. Decryption has to support not just linear replay, but also the so-called trick-modes that emulate the functions of a DVD player: fast-forward, scene-skipping, pause, rewind. This places extra demands on key management and decryption systems.

There are also rights issues. Just because a broadcaster has rights to show a program as part of a linear schedule, with, perhaps, repeat rights, this does not necessarily translate into VoD rights. Back to the lawyers. A third issue is that of navigation. Many people are familiar with the Electronic Program Guide (EPG) grid structure for linear, scheduled TV programming. For VoD a usability redesign is necessary, as the amount of content will be enormously greater, with thousands of titles. The organizing themes can also be diverse—by genre, date, director, actors, and so on.

The Architecture of IPTV Networks

Broadcasters are used to one-to-many distribution networks, often using the phone network as a back-channel for interactive features such as quiz programs, shopping channels, voting, and program selection via pay-per-view.

An IP network provides a personalized two-way broadband channel to each home—potentially to each user. For linear, scheduled TV, this is, however, more of a problem than a feature. If we assume that each channel of standard definition TV can be carried in around 3 Mbps of bandwidth, with reasonable compression, and allowing overhead for ECMs, EMMs, and EPG refresh, then a 300 channel line-up will require around 1 Gbps bandwidth. To minimize the time taken in channel-changing, the preferred architecture is to deliver all the channels to the nearest network point to the customer—the DSLAM or MSAN—and to let that device switch the required channel video signal to the customer.

The easiest way to transfer a dedicated 1 Gbps traffic load from head-end to every DSLAM/MSAN is in the optical domain, where it will not overload the existing IP network routers. Impress the Gigabit signal onto a dedicated wavelength and then use optical multicast to distribute the signal to the required edge nodes. If optical multicast is not available, then layer 2 broadcast could also be used, layering a virtual Ethernet LAN across the network.

Some carriers believe that even so, it is just not cost-effective to carry linear channels across the fixed network (where it incurs significant marginal cost per extra subscriber) when existing radio broadcast solutions can add extra subscribers at virtually zero marginal cost. In this view, the answer is a *hybrid* architecture, in which linear TV is distributed by a broadcast platform, doing what *it* is good at, while VoD is provided by the IP network, doing what *it* is good at.

The Architecture of VoD Networks

The centralized solution envisaged for IPTV's linear scheduling doesn't scale for VoD. Suppose as few as 300 customers sign-up for VOD. Their combined bandwidth is already around 1 Gbps and since each session is temporally, and perhaps content independent, this is bandwidth which has to be provided additively by the IP network.

But, perhaps, not by much of it. It rather depends on where the content is. Putting the VoD servers at a central location will maximize the load on the network. However, as the servers migrate nearer to customers, perhaps at PoPs, then all the traffic is straight from the local VoD server through to the local DSLAM or MSAN and then down the copper wire straight to the customer. The amount of traffic-diversity will be less, this close to the subscribers, permitting smaller video servers while the local caching deloads the central network.

Practically, people envisage a three-level hierarchy. The most popular content, that with the highest probability of being requested, will be pushed as a background task to hard disks within the STB (or PC). With suitable "recommender" automation, this could be personalized. The next tranche of popular titles will be on local cache servers located at or near the carrier PoPs. Finally, an archive set of *video servers of last resort* will be placed centrally. The beauty of this solution is that it simultaneously minimizes network load, and maximizes responsiveness for the customer. Appropriate ordering/subscription, conditional access and billing systems are needed to make it all work.

Triple Play?

It is unlikely that anyone, carrier or broadcaster, would restrict themselves to delivering TV content alone to a broadband subscriber. The up-sale to high-speed Internet access and VoIP is neither difficult nor costly and promises revenues that will more than cover the incremental cost. I say VoIP, but I really mean multimedia session services such as we discussed in the last chapter on IMS. Just because some VoIP is free doesn't mean there isn't a significant and profitable communications business waiting to be built on the basis of upcoming IP session capabilities. To think otherwise is to risk being blind-sided by accidental features of the present situation. We will return to this again in the final chapter.

The Problems with Home Networking

We have not said much about the home situation where the customer is located, but it's there that perhaps one of the biggest problems lurks, the problem of home networking. Unlike phone lines, a DSL connection is terminated in the home at only one DSL modem connection. So how can data be transferred between different rooms (Figure 3.6)? The best solution for data rate, reliability and QoS would be Ethernet cabling, but few householders want to run cat 5 around their homes. WiFi has been the alternative to date, but WiFi can be erratic in practice, has problems with walls and obstructions, and cannot today handle real-time isochronous data like VoIP and streaming media (802.11e will partially address this issue when available). There are also major issues with the small number of

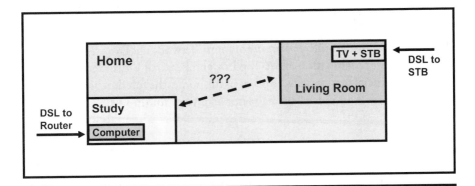

Figure 3.6   Issues in home-networking.

orthogonal channels in 802.11g, and the increased interference if power is stepped up to handle in-building attenuation. Nevertheless, wireless LAN technology *will* improve to the point where it is good enough, the major question is how long this will take.

A promising alternative is to run data along the home wiring, and Ethernet connection plugs are available. Problems with radio interference have been reported, though. Multiple ring mains are also an issue. Whatever home networking technologies are chosen, the issue then arises of self-install vs. technician install. The former narrows the market, while the latter drives up costs and complicates service take-up.

A further issue is the sheer complexity of setting up any kind of network, let alone a standalone home network that requires:

- Configuration of NAT and DHCP services,
- Managing Firewall rules, particularly if working from home (e.g., allowing VPN access),
- Configuring WiFi security—encryption and authentication,
- Providing and maintaining security software such as anti-virus packages,
- Providing operating system and firmware upgrades and patches,
- Systems integration of a diversity of pieces of equipment,
- Troubleshooting.

That part of the addressable market that can perform these functions themselves is just about exhausted. From now on, we are into the "grandmother" part of the market—people who haven't a clue about technology. The only effective solution is a combination of pre-integrated components, initial technician install, and a managed home network remotely overseen by the operator's staff. It will require a good deal of work to provide all of that at competitive prices.

The Opportunity of Home Networking

We should also address the *opportunity* of home networking. From a broadcaster's perspective, the combination of Set-Top Box and Personal Video Recorder (STB-PVR) is the crucial service delivery point in the home, the fulcrum both of lock-in and also up-sell. And indeed there are opportunities for up-sale.

Thin Client

With a TV in every room, there is a requirement for service ubiquity. It would be best if every TV could access the same set of channels, premium services, and VoD. In a thin client model, the main STB-PVR that connects to the DSL line is upgraded to be able to manage multiple encrypted programming streams (multiple tuners). The myriad of separate programming streams are then distributed through the house to thin-client devices attached to other TVs. This provides a lower cost solution, centrally managed within the home, which promises increased revenues to the broadcaster, and presumably increased satisfaction to the TV watchers, each viewing their own favored type of programming in the privacy of their separate rooms.

Plug 'n' Play Device Hub

In a plug 'n' play mode, the STB-PVR hub comes with interfaces (e.g., USB 2.0) into which a variety of consumer electronic devices can be plugged: MP3 players, portable media players, games machines. The STB-PVR can authenticate the devices and assess their capability to receive and play content. It can negotiate specific device capabilities, manage content transfer from the head-end or PVR local cache under DRM control and bill the transfer appropriately. The hub can also manage the consumer devices themselves (e.g., by managing software and firmware upgrades).

PC Emulation

Continuing in the consumer electronics hub mode, the STB-PVR can act as a docking station for up/downloading digital pictures or movies or a printing hub. It can act as a central management console for security web-cams or video-conferencing. It is even possible to imagine connecting work-out equipment to the hub to coordinate dynamic scenery changes with the use of running or biking home-exercise platforms (NDS 2005).

However, it has to be said that the PC industry and the consumer electronics industry both have their eyes firmly set on dominance in the home networking multimedia space. The eventual winner, if any, is not yet apparent.

Summary

We have covered a great deal in this chapter. We started by looking at the TV industry value chain, based today on linear channels, and discussed why channels are bundled. We noted that the arrival of the Internet has removed the broadcast channel bottleneck, and made it possible both for content owners to potentially disintermediate established broadcasters, and to bring to market their "long tail" of inventory at acceptable, and even very low, costs. We looked at pricing models for "long tail" content offers.

Next we looked at some of the non-TV opportunities enabled by the Internet. These included both gaming and new kinds of services. The technologies are mostly here, or are about to arrive. What is needed is platform integration and productization: both are tasks that the larger carriers have the resources to accomplish over the next few years.

Then we turned to the specifics of broadcast infrastructure technology, which enables linear channel IPTV, Video-on-Demand, and their conditional access systems. We looked specifically at how to implement IPTV and VoD on IP networks, and the trade-off between caching information on servers, and transmitting it through the network. Finally, we looked at the triple play options and the difficulties and opportunities of bringing all these services to reality in the home networking environment.

The issues discussed here will be revisited in the final chapter, when we assess the strategies being adopted by each of the major types of player.

References

Broadcasting Policy Group. 2004. Beyond the charter—the BBC after 2006, February 2004. http://www.beyondthecharter.com/.

International Game Developers Association. 2005. Casual games White Paper. http://www.igda.org/casual/IGDA_CasualGames_Whitepaper_2005.pdf.

NDS. 2005. *The NDS guide to personal tv.* Middlesex, UK: NDS Ltd.

Rollings, A. and Morris, D. 2004. *Game architecture and design,* pp. 409–420. Indianapolis, IN: New Riders Publishing.

Tranter, S. 2004. Conditional access, simulcrypt and encryption systems. In *Broadcast engineer's reference book.* E. P. J. Tozer, ed., 385–394. New York: Elsevier.

Wray, R. E., John E. Laird, Andrew Nuxoll, Devvan Stokes, and Alex Kerfort. 2005. Synthetic adversaries for urban combat training, *AI Magazine,* Vol 26, No. 3, 2005. http://ai.eecs.umich.edu/people/laird/papers/wray-2004-IAAI-MOUT.pdf.

Chapter 4

The Next-Generation Network and IT Systems

Introduction

In Information Technology circles, the discussion is all about Service-Oriented Architecture, Web Services, and Grid Computing. A few forward-looking individuals even know a little about Web 2.0 and the Semantic Web. But take even a cursory look at the IT systems inventory of any established carrier and you will find systems going back to the 1980s and earlier. Carrier systems are rarely retired: instead, new applications and architectures overlay old ones until a kind of geological stratification occurs.

The problem with carrier IT is not so much how to absorb new technology, as to figure out how to use it cost-effectively to deal with the legacy of the past: both the obsolescence of ancient hardware, operating systems, computer languages, and applications; and the spaghetti of standalone systems, ad hoc interfaces, and manual workarounds, and re-keying of data. We could also include the lack of any common data architecture, schema, naming scheme, or record format consistency.

In carriers, we normally distinguish between BSS (Business Support Systems) and OSS (Operations Support Systems). You would find BSS in any large enterprise: these are the systems that support standard business processes such as sales and marketing (CRM), enterprise resource planning and management (ERP/ERM), and billing. OSS is much more tightly focused to telecoms, including the element and network level management systems that configure the boxes, acquire and aggregate status information, and manage faults (trouble-ticketing).

I will start with a diagnostic tour around a typical carrier's IT infrastructure, and its attempts to move forwards. Although this is based largely on my own experience, please be assured that what is about to be described is completely typical.

The State of Carrier BSS and OSS

Towards the end of the Internet boom, I was appointed chief architect at a global carrier, with particular responsibilities for information technology systems. During the 1980s and 1990s, the carrier had steadily accreted systems. It sometimes seemed that in the three dimensional space comprised of products, processes and networks, every resultant cell had its own, special IT application to make something happen. Perhaps you suspect exaggeration? Figure 4.1 shows the UK Business Support Systems (BSS) map across products and processes at that time.

Products are listed down the left-hand side. Business processes are listed along the top. The boxes are IT systems. To avoid any confidentiality issues, IT system identifying information has been removed and affine transforms applied. Each box is a system, and 90 percent of the boxes are separate systems (rather than the same system supporting different functions).

The problem is worse still. I have not shown the Operations Support Systems (OSS) used to monitor, provision, configure, and control network equipment, and to handle alarms. The corresponding OSS diagram contains a further constellation of IT systems amounting to more than half the number of applications documented in Figure 4.1.

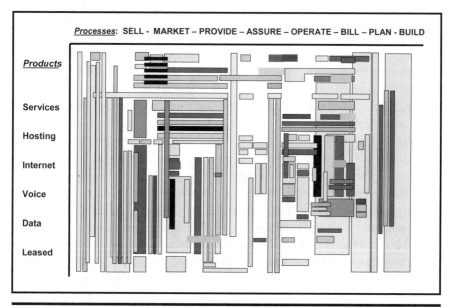

Figure 4.1   A typical carrier systems map.

Nor were the applications as decoupled as the diagram suggests. Some key applications (such as network inventory management applications) had as many as 40 to 50 interfaces into other applications that needed to use them. These interfaces were entirely custom-built, and had been added over many years in a variety of computer languages and networking paradigms. This proliferation of diverse, complex systems with, in many cases, manual interconnect (requiring data re-keying) had many negative consequences.

- Processes were locked down by the inflexible computer systems.
- Due to the detailed product-specific implementations, systems and process re-use was very difficult. To add a new product one was faced with the unenviable dilemma of either expensively introducing a new suite of IT applications, or engaging in a very expensive rewrite and customization of existing applications. Some of those applications were written in Fortran.
- There were no economies of scale either in processes or systems.
- Even small changes had to go into the IT pipeline to be fixed. This frequently took months. Some departments viewed this as a lack of responsiveness and took to recruiting their own secret groups of developers, outside IT's view or control. There was an IT underground of rogue development and applications.
- IT was generally hated and despised, and became a scapegoat for program failures. This last point particularly irked the IT staff.

I am not being particularly harsh about my employer at that time—all carriers with some history behind them have exactly the same problems. And it was not as if we didn't know how to design problem-solving IT systems. The industry had wrestled with inflexible, stovepipe IT throughout the eighties. A consensus had developed around the following approach shown in Figure 4.2:

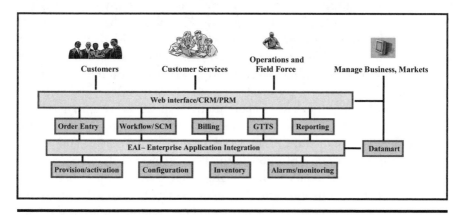

Figure 4.2   Today's BSS-OSS architecture.

- Use COTS (Commercial Off-The-Shelf) packages, rather than in-house applications,
- Use enterprise application integration middleware to "glue" applications together,
- Use Internet technologies for maximum flexibility.

In fact, one of my counterparts from another carrier had evangelized a similar target architecture around his organization, under the name of "'the star-ship"—there *is* a similarity if you look carefully.

For a "green field" carrier (i.e., a start-up), there were no issues. Just do it like the star-ship. When we acquired a major U.S. data hosting company, all of three years old, we found it had a modern flow-through automation layer exactly aligned to the model, and sourced entirely from PeopleSoft (now Oracle). We spent months trying to decide if we should break their model by replacing their PeopleSoft financial component with SAP, as used by our extremely influential finance people.

For "legacy carriers," however, structured entirely around incompatible architectures and systems, and unable to be shut down for the purposes of migration, the issue is how to introduce new architecture and technology *at all*.

The consensus to-date has been to migrate from in-house applications to COTS packages and to use EAI —Enterprise Application Integration—to tie applications together. EAI is the magic ingredient here. The EAI application, or hub, supports standard interfaces to standard applications—it can exchange messages with Siebel, SAP, popular billing applications, and so forth. It then acts as a post office, accepting messages from one application and delivering them to another, according to configurable business logic scripts. EAI interfaces can also be written for a carrier's proprietary legacy applications, although usually at a cost of millions of dollars. The currently fashionable phrase is to call this architectural concept the Enterprise Service Bus, especially when it supports web services interfaces.

For the last few years, we have been meant to get excited about web services as the real answer to systems modernization. In the web services model, applications are like "objects" with public interfaces—"methods"—that can be invoked to execute a business or network function. Applications publish their web services public interfaces in a UDDI registry (UDDI = Universal Description, Discovery, and Integration) that can then be searched by other applications in a "yellow pages" model.

It is not sufficient just to have all applications promiscuously exposing random interfaces in case someone else wants to access some of their functionality—the exercise must be structured and controlled. A Service Oriented Architecture (SOA) is one that specifies needed functionality and determines which interfaces ought to be published. Legacy applications can also be given web services interfaces (at a price!) so that they can also participate in the Service Oriented Architecture.

Within a web services world, the distinction between the application and the EAI middleware gets eroded. Applications are designed from the very beginning to import and export their specific functionality and communicate bilaterally. It looks more like a meshed world than a hubbed one.

How does an application talk to another using web services? Recall that the web services model is object-oriented, so that a message is sent to a web services interface, and after a computation, a reply message is sent back. This is like method invocation in an object-oriented programming language (e.g., Java) or a procedure call for any reader whose last experience of programming, like the author's, was COBOL. The procedure call is coded in an XML format, and inserted into a SOAP message. (The function of SOAP is to identify the method, package the parameters, set-up a transaction-id to coordinate the request and response to manage security encryption/decryption and to handle any other transaction-related administration—it is like a combination of envelope and routing note packaging a document).

For example, suppose that an application connected to a music download site needs to check with the inventory system of a supplier. From its own database, the application believes the product ID is PQ85a. It therefore sends a SOAP message to the inventory system asking for full product details as shown in Listing 4.1.

```
<soap:Envelope xmlns:soap="http://schemas.xmlsoap.org/soap/envelope/">
  <soap:Body>
    <getProductInfo xmlns="http://catalogue.example.com/dl">
      <productID>PQ85a</productID>
    </getProductInfo>
  </soap:Body>
</soap:Envelope>
```

Listing 4.1   An example SOAP message request.

The inventory system now does a look-up in its own database and finds the required product details, which it sends back in the response shown in Listing 4.2 (adapted from the Wikipedia discussion of SOAP).

```
<soap:Envelope xmlns:soap="http://schemas.xmlsoap.org/soap/envelope/">
  <soap:Body>
    <getProductInfoResponse xmlns="http://catalogue.example.com/dl">
      <getProductInfoResult>
        <Category>Classical</Category>
        <productID>PQ85a</productID>
        <description>Goldberg Variations</description>
        <Performer>Glenn Gould</Performer>
        <Composer>J. S. Bach</Composer>
        <price>20.00</price>
      </getProductInfoResult>
    </getProductInfoResponse>
  </soap:Body>
</soap:Envelope>
```

Listing 4.2   An example SOAP message response.

SOAP is normally carried across networks using HTTP, just like HTML messages between browsers and Web sites. As can be seen, SOAP messages occupy a lot of bytes, and routing messages around networks also appears somewhat challenging. In fact web services has opened up a new application-level network layer, centered around handling SOAP and XML messages that the existing EAI hub vendors have been quick to exploit, along with traditional network equipment suppliers.

Functions at this layer include access management and security, routing of messages, schema transformation, performance monitoring, load balancing, compression, and caching. Devices supporting such functions in hardware or software are called "content-aware network appliances." They can be folded into blades running on network switches and routers, as in Cisco's Application-Oriented Networking (AON). IBM, Intel, and Citrix are also prominent players. This level of attention has also identified some weaknesses in the existing web services architecture: specifically the current absence of standardized application-level routing protocols between SOAP end-points, paralleling the services provided by IP routing protocols such as OSPF.

I mentioned we were meant to get excited about web services, but over the last few years the impact of web services on carrier IT has been bounded, to say the least. Web Services started as a concept, then developed into an architecture and moved to standardization. Only then were programming language interfaces and system development kits able to emerge, and their first versions were lamentable. We now have stable and sufficiently sophisticated web services development platforms, but it is taking time to re-engineer traditional COTS packages into the new architecture. Until that is done, re-tooling carriers will continue to be difficult, even ignoring the issues of legacy. And this is the reason why the revolution is progressing on the timescale of a decade, rather than months or years.

Project Ultimate and Project Diamond

Shortly after my arrival, I was introduced to the showpiece IT project that was to save the company. I will call it "Project Diamond." Project Diamond was the *second* attempt to sort out the log-jam of obsolete systems that had dragged my employer into the morass. Before Project Diamond there had been an initiative that I will call "Project Ultimate."

Project *Ultimate* had been the ultimate big bang. The CIO had frozen all IT spend and had apparently removed his team and himself from our own universe to design and implement Ultimate in great secrecy somewhere else. Ultimate was not conceptually wrong; its architecture was exactly that of Figure 4.2. The problem with Ultimate was that six months had gone by, all departments had backlogs of work that IT had refused to work on, new products could not be introduced, and when finally the great Ultimate launch meeting was held, no one in the audience could understand what on earth was being proposed.

After the CIO had duly relinquished his post, big bang transitions became unfashionable. The new incumbent decided on an incremental approach, hence project Diamond. Diamond was also an instance of the target carrier architecture of Figure 4.2, but rather than trying to replace all of the existing IT systems at once, it had the incomparably simpler task of just supporting a couple of new products. Even legacy data migration would not be necessary.

The people doing Diamond were from a major consultancy, and you could not have met a nicer bunch. They were friendly, hard working, and frequently took us to dinner. They made all the decisions themselves, and integrated sensible mainline applications (not all of which were in use in our organization at the time). Diamond delivered a textbook system, roughly on time and at staggering cost. It didn't take long before we noticed that no one was using it. For Diamond to be inserted into the life of the organization, major products would have to be migrated onto it. This meant that the elegant end-to-end automation system the consultants had built would have to be integrated with tens to hundreds of legacy systems:

- Network inventories to check circuit availability,
- Customer databases for order management,
- Billing and invoicing systems (of which there were many),
- Existing fault management systems,
- Partner management systems.

And, of course, there were many different systems under each of these general headings across the different products and geographies. As an integrated, flow through automation platform, Diamond could not deliver its value until these interfaces were in place.

The consultants were prepared to grit their teeth, get stuck into what was evidently shaping up to be a multi-year project, and continue spending our money. Meanwhile, products were being further delayed because they were told not to deploy on legacy systems, but to wait for Diamond; the routine maintenance that every department needs was still not being done. Tensions mounted and no IT job looked safe.

Soon we had a new plan. Get the front-end CRM process sorted out and we could at least get the benefits of a standard package in use across the organization. Flow-through automation would have to wait. I believe this was the first point that reality had intruded into IT for at least a year. Of course, because we were now confronting real departments with real customers and real processes, we started encountering real problems. Locally, the front end-processes, although idiosyncratic, were not bad. They had adjusted to failings in downstream systems and over the years had been customized and tinkered with until all the internal departments were happy. The new, standard system had none of these idiosyncrasies of course. It reflected the vendor's model of some generic industry norm for doing CRM and order entry, and like a new shoe, it pinched. But it did have one major advantage over the status quo—superior, automated reporting to senior sales management. This was enough to get it steam-rolled into operation.

As business conditions worsened, the consultants were let go and development reverted to in-house staff. Slowly the grand transformation plans were abandoned, and incremental development became the norm, focused on where the pain was greatest. Strategic progress stopped altogether, but surprisingly, at the margin, where the pain was the worst, things improved.

For example, every month the sales and operational senior managers got together to plan production for the next period. What they wanted to do was simple: review the orders pipeline and network occupancy so that they could schedule capital spending in advance. The idea was, as far as possible, to do just-in-time investment. Of course, the IT systems were incapable of delivering this information. Customer and order information was shredded across dozens of databases and spreadsheet files. The network capacity information was also spread around amongst transmission circuit databases, switch loading statistics and Frame Relay and ATM virtual circuit files mapped to switch occupancy figures.

The COO set up a small team to fix this problem, ignoring IT altogether. The task force identified the relevant databases (there were more than 40) and wrote Web scripts in ColdFusion to automate data retrieval. In the case of spreadsheets, they had to write tailored code to map these into a database.

Every month they would access the databases, pull out the information, reformat it, combine and summarizes it, and then produce formatted Web pages and RAG (Red-Amber-Green) reports. Sticking plaster, of course, it was, but I attended some of the sales-operations review meetings, and they would not have happened without it.

In the end we had teams of Web-engineering programmers who could glue Web front-ends onto legacy systems and provide a superficial layer of integration to grateful staff. It was useful. Meanwhile, the heavy engineering of COTS introduction and legacy dismantling went on in the background, to a muted bass tone of user pain, and the gurgling sounds of millions of dollars seeping away.

Is There a Better Way?

The CIO job in a carrier old enough to have layers of legacy systems is a completely poisoned chalice. The CIO and his staff are often some of the brightest people around. These days, they are as well versed in business realities as in the technical state-of-the-art. Yet change doesn't happen, and careers are destroyed. Why?

Consider Machiavelli's overfamiliar observation from chapter 6 of *The Prince.*

> And it ought to be remembered that there is nothing more difficult to take in hand, more perilous to conduct, or more uncertain in its success, then to take the lead in the introduction of a new order of things. Because the innovator has for enemies all those who have done well under the old conditions, and lukewarm defenders in those who may do well under the new.

A modern IT architecture is indeed a "new order of things" and it needs a number of separate transformations all to be successfully accomplished, with benefits accruing mostly at the end.

The task has to be modularized—a big-bang cannot work. But if only some of the processes and systems are to be modernized, interfaces have to be built to the remaining legacy systems, otherwise the process-flow breaks. How can such expensive new interfaces be justified when they only connect to ancient systems that are due to be phased out in their turn anyway?

It will be found that data is shredded across the organization. I have already alluded to the fact that information about sales to particular customers was spread across 30–40 different databases. And some of these databases were spreadsheets. And the field names were different in the various systems. So it was never quite clear whether the reference was to the same customer and product, and whether we were double counting.

There is no one to do the work. Most people today in carrier organizations are working many extra hours per week just to do their line jobs. There is no slack to document existing processes, design new ones, configure and tailor the COTS packages from out of the box, run the necessary trials and implement pilot programs, to educate the staff and to manage the inevitable headcount reductions. It is an uncomfortable truth that most of the desired OPEX improvement resulting from major IT investment comes from the headcount reductions achieved through more efficient automated processes.

I do not know of a single case of a legacy carrier successfully migrating its entire organization to a successful, state-of-the-art IT system, together with a suite of flow-through processes. As we found from our data center acquisition, there have been cases of start-ups, particularly in the Internet boom, who were able to build their IT from scratch—the resulting efficiencies were proof of the virtues of success.

Does this mean that it is hopeless? Almost. Only the gravest crisis can justify the expense and disruption of a comprehensive step-change in processes and systems. The transition to the next-generation network demands sufficient product and network innovation to cause just such a crisis. However, it is still so expensive that an undercapitalized alternative operator may well be unable to meet the bill. The NGN will be the catalyst for a wave of consolidation leading to healthier businesses able to charge premium prices and capable of investing in IT renewal.

A Common Data Model?

One of my few successes as chief architect was my cancellation of the Common Data Model project. The consultants concerned had managed to convince a number of senior managers that what was needed was a common data model across the organisation. They had prepared a large number of binders when I first encountered the project, detailing the theory of data modeling, class and inheritance models, modeling languages and standardization efforts. They were all set to spend further millions of dollars walking around our organization documenting the myriads of names different projects had come up with over the years for what we ought to recognize as the same real-world entity. I went ballistic.

What was wrong was that the consultants had no answer to the question what would change if tomorrow you gave us a completed common data model; what would we do with it? There was no possible answer, because in the absence of standardized databases and applications adhering to the model, a common data model was just so much useless paper, obsolescing by the second. We had no IT program at all that could be plugged into the Common Data Model activity, and what's more, the consultants knew it—easy money.

I don't believe that for large carriers a single, complete, and consistent data model is possible, although it is a worthy objective. There are just too many different processes and applications, with different data schema requirements. The key thing is to get the data modeling process under control. There must be a schema management procedure in the IT department—part of the gate process by which new IT applications are released to service. Today that means an XML-based schema, and it is possible to translate between different, compatible schema using standard XML tools, linked with the messaging middleware. This kind of distributed scheme is scalable, flexible, and robust against new corporate acquisitions.

At time of writing, I read that IBM is pushing Master Data Management (MDM) (*Information Age 2006*). Apparently the centralized MDM database should be updated constantly, so that it can push altered data to applications to ensure everything is always in synch. And of course, this needs a high-bandwidth network and a Service-Oriented Architecture. Naturally, getting to this Nirvana cannot be a big-bang, but must be a multi-year process. Still, Gartner apparently conclude that by 2010, over 70 percent of Fortune 1000 companies will have implemented MDM programs. It coyly does not speculate as to whether any of these programs will have succeeded, or what the other 30 percent thought they were doing.

A Correct Strategy for IT Transformation

Having described a number of cases where change didn't work, are there any lessons for success? I believe the correct strategy for IT transformation is simply this.

Realize that IT transformation is simply an enabling mechanism for *business transformation* to a new, more efficient and lower-cost business. First commit to the business transformation program, then commit to the IT modernization program as a key enabler.

No business transformation plus IT evolution can be done purely with internal resources. It needs a resourced program team with a mandate to engage with the entire organization and *turn it around*.

No business case can be made for transitioning the entire organization to the IT state-of-the-art. An audit will show that certain products (such as circuit-switched voice and synchronous transmission) are probably not too bad, although based on obsolete processes and IT, and consequently with a too-high cost base. But since we are going to close down these products and networks anyway as soon as we migrate to IP-based products, the right thing to do is maintenance only: apply process and IT *sticking plaster* where the extra revenue from any marginal improvement clearly and immediately covers the marginal cost.

Finally, there are so many vested interests who seek to slow down, avoid and dilute process and systems change that success will only occur if led from the top. The CEO and C-level managers must not assume this can be delegated down the line. They need to chair the program board, and it needs to meet often.

In the nutshell, carriers have not been good at reinventing themselves, and that is why their IT transformation programs have failed.

An Internet Self-Service Model

Anyone who has used a company with an Internet sales channel like Amazon, Cisco, or Dell will ask themselves why carriers cannot provide their customers with a similar interface. I have heard the excuse that the carrier business has a far more complex product line than the relatively simple product models of online retailers. Agreed, a book is not too complex, but Cisco routers and Dell PCs? Please!

The real answer is that an E-business architecture turns the existing carrier waterfall-type IT architecture inside out. In the carrier IT systems model we usually see today, the customer and sales person get together to configure and price the order, which then enters the front-end CRM system. The order then burrows deeper and deeper into the BSS and OSS , down to the network boxes, out to the field force and across to other carriers (e.g., for buying in private circuits). Deeper in again, the order percolates through to the billing system and various management information systems. Finally, by some magic, tens or hundreds of subtasks run to completion, often after multiple resubmissions to correct errors,

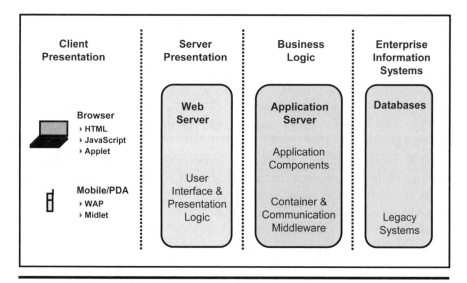

Figure 4.3 Three tier architecture.

across tens or hundreds of separate systems. The service finally becomes available to the customer. Perhaps more than 60 days have gone by. It's hard enough for the carrier's staff to find out what is going on. As for the customer—not a chance.

How do we get from there to an Amazon point where we appear to be able to point our browser at "My Account" and find out everything—what we ordered, where it is, when it will arrive, all our billing and shipping history, our account and transaction history?

The answer is a customer-centric IT architecture organized around central integrated applications and databases. Technically, this is a three-tier architecture (Figure 4.3) generalized to n-tier architectures, as further layers of processing differentiate from this simple model.

In a three-tier architecture, the customer's browser is pointed at the top-tier, the Internet portal site. The portal's technical function is to receive HTTP requests from the user, and to return HTML back to the browser, but a portal does a lot more than that. As the service gateway, it proves the first level of service integration. Different parts of the page serve-up different functionality to the customer—windows to diverse functionality encompassing the totality of the underlying process network.

The middle-tier is the domain of application servers, running business logic components. These are the present form of what we used to call applications, and a coherent set of business functions could be layered onto a number of physically separated but networked components, running on different servers in different data centers. Communication is mediated by web services and SOAP messages.

The third tier comprises databases, or those legacy enterprise information systems not yet assimilated to this architecture. A business component will treat the third tier as an information resource.

Familiar instantiations of the three-tier architecture include the Java-based J2EE, and Microsoft's .NET.

The power of the three-tier concept is the separation of concerns. The user experience can be designed autonomously within the top-tier portal. Application logic can be developed without regard to data formats in the middle-tier, which is where enterprise application integration is also addressed, while all the issues of data format, schema development and data modeling, as well as legacy system integration are concerns that can be addressed within the third tier. It has been a real stretch, however, for carriers to adopt this model as fundamental to their IT.

As already observed, vendors of COTS packages are still re-engineering their historically monolithic applications to this new model, while interfacing legacy applications into the tier 3 position of back-end enterprise information systems and databases is a hard, grinding, and expensive uphill task. But perhaps the most profound reason for inertia is that the dominant "factory" model of carrier processes is not customer-centric at all. Instead, it is optimised around a division of labor and work-handover within its interior bureaucracy. Pervasive application integration can undermine the current waterfall process-model paradigm, but as the old joke has it, to adopt the new three tier, customer-centric architecture, you would have to really *want* to change (cf. how many psychiatrists does it take to change a light bulb? Just one, but the light bulb really has to want to change).

To summarize: the three-tier/n-tier architecture is the right one for today and the only way to achieve reasonable customer service but it has to be *consciously* adopted. The dangers of upgrading to a rather traditional linear COTS + EAI model that has not been modernized as per the Internet architecture above should be obvious, even if it seems easier and feels more comfortable.

Grid Computing

After web services and SOA, I would now like to turn attention to the third major innovation area in IT today, Grid Computing. I first heard about Grid Computing in 2003 at an IBM conference in Florida. Our ever-attentive IBM account manager took great pains to bring it to our attention. I didn't understand it at all. What was Grid Computing? Was it something to do with autonomic computing, the somewhat unconvincing image of a self-configuring, self-managing, and self-healing computer infrastructure that IBM was pushing at the time? Was it some kind of variant of web services? Was it a new IBM operating system?

Even as we grappled with the concept, we were not clear what problem it was trying to solve, or whether we actually had that problem. A few years down the track, and the answers to these questions are becoming a little clearer. Enthusiasts like to explain that the idea of grid computing is that computer power should be a utility, like water or electric power. Just plug your terminal into the computing grid and extract enough computing power to solve your problem. Obviously, there would have to be excess computing power available that a used could tap into with the usual concerns for security and cost-efficiency. This seems to me, though, to be the *least likely* application of Grid Computing. Here are some more plausible scenarios:

- You are an enterprise with a diversity of computing resources across your data centers and desktops. You have little idea how CPU, memory, and disk resources are being utilized, as you can neither measure them, nor systematically allocate them to your application schedule. But you are sure that your usage is grossly inefficient.
- You are a collaboration between a number of organizations with strong joint computing requirements. You would like a uniform virtual computing environment that allows anyone to submit jobs and get the results, but this has to be implemented across half a dozen different IT organizations with different processes, approvals, security, and standards.
- You have invested in a data centre and servers, and you would like to make a business of selling processing to all-comers across the Internet. If only there was a way to set-up, package, and bill for this service in a secure way. (Perhaps this last scenario is not unlike the naive utility model mentioned above.)

To make any of these scenarios work, what is needed is a kind of distributed operating system that can sit on top of a scalable network of distributed computers. Looking down, this new software has to know about the machines it has enrolled, and track their utilization and capabilities. Looking up, it has to accept jobs from clients, map them optimally onto machines with capacity, monitor execution, handle exceptions, produce billing records, and manage security. Welcome to the world of Grid Computing.

The Global Grid Forum

The world-wide efforts to put grid computing onto the map are being led by the Global Grid Forum (http://www.ggf.org/). According to its Web site, the GGF represents more than 400 organizations and is the leading global standards organization in Grid Computing. A tutorial on Grid Computing is available at CERN, the particle physics research establishment where the Web was originally developed (http://gridcafe.web.cern.ch/gridcafe/). The Global Grid Forum was formally set up in 2001, and organizes conferences, supports working groups and produces technical papers. It sponsors the Open Grid Services Architecture (OGSA), the global standard for Grid Computing.

Grid infrastructure reuses many of the standards and interfaces of web services (for example, for security). The major extension required for grid computing is the management of state. While web services are stateless, grid computing has to explicitly manage the state of computer resources. It has therefore introduced the WS-Resource standard.

Many organizations experimenting with early grid implementations are using the Globus Toolkit, (http://www.globus.org/) that implements the OGSA. At the time of writing, the Globus Toolkit is at version 4.x. and has received excellent reviews. For more information on Grid Computing see Taylor 2005.

Using a Grid Infrastructure

Like any operating system, a grid infrastructure will manage secure access, and allocate permitted jobs to available resources. Functions in existing distributed middleware architectures such as CORBA—distributed fault-tolerance and parallelism of execution, for example, will also be supported. The utility of grid computing in enterprise data centers is obvious. Whether it will permit existing unused desktop computing resources to be brought into the virtual server space is less clear—the IT staff have far less control over these machines, even as to whether they are switched on or not. Desktop resources have also to be shared with the allocated user, who may well object if the system is slowed through executing high-priority back-office functions.

Grid computing may make extranets easier to set up, but despite the emphasis on virtual organizations in the grid computing literature, commercial applications still seem a stretch. Most of the existing work is with large scale "big science" projects. Likewise, the concept of a public Internet grid, where jobs can be executed on a networked cluster for a fee, seems a long way off. Most applications run perfectly well either on the client machine, or on an enterprise cluster that seems to the user to be a remote enterprise machine accessed via the browser and HTTP (even though it could well be executing on an enterprise grid in implementation). Enterprise grids are being promoted by the Enterprise Grid Alliance (http://www.gridalliance.org/), but progress still seems slow.

So the conclusion on grid computing is that it will certainly do for distributed computing infrastructure what web services aims to do for enterprise application integration. And like web services, it will probably be a while in arriving in a form that enterprises and carriers would actually want to deploy internally.

Web 2.0

Considering this chapter so far, I am rather aware of "big systems chauvinism." Much discussion about back-end systems, not much about users, particularly carrier staff. There is a category of IT called "desktop support." This area deals with PCs and horizontal productivity applications such as e-mail and runs the IT help-desk. Although it will offend those working in this area, I have to say that, from a CIO perspective, this whole area is not very exciting. When it works well, nobody cares, and when it breaks, it's nothing but grief.

Part of the reason for the lack of excitement is the plateau of functionality. We seem to have got to Microsoft Office quite early, perhaps in the mid-nineties. That gave us e-mail, contacts and calendar, a spreadsheet, word-processor, and a presentations package. Since then, possibly excepting tools such as diagramming systems for engineers and desktop databases for analysts, it has been hard to identify any new, truly pervasive horizontal applications.

This has been frustrating for people who wanted to move up a gear into collaborative tools. The Intranet has managed to move some traditionally paper-based forms onto screens, but seems in general to have disappointed. It is hard for staff to publish to enterprise Intranets, and hard to find stuff once it's there. But the Internet continues to be a source of technologies both scalable and usable. It's time to think about Web 2.0.

A few years back, I was briefly enthusiastic about the semantic Web. This continues to be a multi-year effort from the world-wide Web consortium to make Web content semantically clean and precise, and independent of any underlying application. The idea is that when you search on, for example, the word 'rock', the resources out there on the Internet should know whether they pertain to geology, music, an actor or repetitive movement. To do this resources need to be described in a language that is both rich enough to capture distinctions and relationships and is equipped with a broad enough, standardized vocabulary—an ontology—to specify the type of thing resources are, and the conceptual class hierarchies to which resources belong.

I recall attending a meeting about the Semantic Web with Professor James Hendler, a leading researcher, back in 2002 at Virginia's Center for Innovative Technology. "Who exactly is going to tag the millions of Web pages with semantic Web mark-up?" I asked. Professor Hendler was not sure, and I thought at the time this was a bit of a show-stopper. If people weren't going to do it, then natural language machine systems were going to have to get a whole lot smarter in understanding content (see chapter 12). And if they became that smart, perhaps we didn't need to do the semantic Web tagging in the first place?

Five years later and the Semantic Web is still work-in-progress, but the idea of tagging has caught on. Not the finely analyzed conceptual hierarchies of the professional ontology developers, but folk-tagging by millions of ordinary users applied to user-generated content on an increasingly large number of sites. There is a widespread belief that there are powerful network effects that will drive tag-convergence to a standard folk ontology from below—we shall see.

User-generated content sites that encourage tagging need to provide automation to assemble ontologies bottom-up from the most popular tags being proffered, and to present possible tags to their customers as they seek to classify. This is the fast way to get convergence. With widespread and consensual tagging in place, the synergy with search engines is readily apparent. The promise is 80 percent of semantic Web functionality for 20 percent of the effort.

Another not-wholly-predicted success story has been wikis, such as the Wikipedia. Again, the software provides an easy to use framework for users to add and review content—their own and that of others. Arguably-utopian beliefs that convergence to a high standard of content would occur seem to have been largely born out in practice, lubricated by a light touch of moderation (helped along by moderators).

Some of the more enterprising vendors are putting all these technologies together as enterprise wikis with tagging and search. For enterprises whose main asset is the intellectual property of their staff, this is probably a new kind of desktop software worth having, once past release 1.0. I would like to believe that next-generation network carriers should also be included in this elite category.

Web 2.0 should not be *conceptualized* in terms of its new applications (wikis, etc.), or its user-generated content, or its software-as-a-service potential. In today's Internet, connectivity is important, of course, but the *paradigm* has been defined by specific applications: Web servers, e-mail, and so forth. In Web 2.0, the paradigm is *connectivity itself*. Around the notion of platform connectedness we see an explosion of new ways to converse: asymmetrically and peer-to-peer, people and application systems. As we discussed in chapter 2, the NGN is foremost a middleware platform, a nursery for new patterns of communication, which can then be packaged as services.

Summary

In this chapter, we have looked at the train wreck that is today's legacy IT infrastructure. The result is strategic immobility, as it is usually cheaper to build an additional IT layer alongside existing systems for a new product. The overall ratcheting-up of costs due to this added complexity is not usually identified in the business case.

We then looked at modern ways to build carrier IT systems, using the "Enterprise Service Bus" concept and a "Service-Oriented Architecture" exploiting web services.

The practical problems of IT transformation were then examined in case studies of projects Ultimate and Diamond. We briefly touched on issues of data management, and a common data model. We then looked at strategies for transformation that have a greater prospect of success, and explored an Internet self-service model for carrier IT.

We finally assessed Grid Computing and Web 2.0, both flagged as new technologies and concepts that need to inform the next-generation network project..

Getting back to basics, the critical problem, as always, is that of legacy. Legacy systems cost more to adapt to future methods of operation and new systems architectures than they are worth, and strangle attempts at process efficiencies. The beginnings of wisdom lie in the recognition that most legacy processes and systems are unrecoverable, and must be left to wither. The NGN is an opportunity to start anew, and the challenge for top management is to grasp this point, and force through the next-generation organization against all resistance. The consequences of half-measures will be bleak indeed.

References

Machiavelli, N. 2005. *The prince.* New York: Oxford University Press. http://www.constitution.org/mac/prince06.htm.

The Master Key. *Information Age*, p. R3, March 2006. (IBM sponsored supplement). http://www.infoconomy.com/_data/assets/pdf_file/75325/RR10-LR.pdf.

Taylor, I. J. 2005. *From P2P to web services and grids.* New York: Springer.

TRANSFORMATION II

Chapter 5

Bureaucracy and Treacle

Driving Change in a Carrier

Carriers have an image problem: they are widely perceived to be bureaucratic, slow-moving monoliths. The apocryphal sales pitch "Buy from us, we suck less" (than our competitors) was attributed to an executive from a North American carrier. Sadly, only the better operators could plausibly make that claim.

Senior telecoms executives share this perception. They see the organizations they are meant to be leading as opaque and unresponsive. Their failure to drive change often results in a regular turnover of senior, accountable staff. We used to joke that new CEOs were like the pilots of huge jets flying across the Atlantic. Everything is fine while the plane has to fly straight and level, but as it enters a region where it has to manoeuvre, the pilot discovers to his horror that the controls do not, in fact, connect to the aircraft.

All organizations have a tendency to bureaucratize. An organization's ability to function is grounded in its processes, which should be independent of the idiosyncrasies of the people concerned. Good processes represent the intelligence of the organization—it's often said that good processes are the way we get superior performance from average people. As organizations get bigger, processes solve the coordination problem between multiple agents across different times, geographies, and skill sets. Processes are partially implemented in computer applications and telecoms networking: the rest comes down to people playing the roles the process stipulates.

Processes normally evolve in a bottom-up and incremental way. Some new situation arises that the existing process cannot handle; the people most concerned fix it using the easiest means to hand. The way problems are solved in organizations is by way of projects—smaller projects are faster and easier to get approved than larger ones. So processes most often change by small fixes.

Like the similar evolution of software code, the result of years of incremental process maintenance is a heap of *process spaghetti*—arcane rituals that make no sense to the outsider, obsolete methods that could easily be replaced by something much more efficient ("it ain't broke, so why fix it?" say the staff). When scaling a process, it is almost always easier and pleasanter to add some more people than to fundamentally improve the process and thereby disrupt things. It is in few people's interest ever to remove staff subsequently, which is why mature processes have a bloated headcount. Often there are so many people involved in running processes that inter-process coordination becomes a significant internal problem, requiring the construction of further meta-processes—steering boards, coordination committees, internal public relations and communications staff, internal client managers. More headcount and further impediments to change in the resulting "veto network."

Being bureaucratic and immobile creates market unresponsiveness: the input-process-output loop grinds slowly due to internal friction. This looks to customers like stupidity, and opens a market opportunity to nimbler competitors who can give the market what it wants. This, of course, assumes the market is deregulated, and that the barriers to entry are not too severe. If competitive pressures matter to the bureaucratic organization, it will respond to lost orders and falling revenues by reviewing its internal processes and attempting drastic surgery. At best, this *could* amount to major process-re-engineering, and new cycles of automation. A volatile commercial environment and the fear of competition keeps companies lean and responsive. As soon as the pressure is off, bureaucratization starts again.

Telecoms Market Structure

The telecoms sector has not historically faced the levels of competition that other industries have experienced. During most of its first hundred years, telecoms was considered a natural monopoly and was regulated as a utility (regulation is not normally very effective in combating bureaucratization).

When the telecoms sector was liberalized, the structure changed to an oligopoly. Oligopoly is a market structure where there are only a few players (2–8, typically 3–5) and significant barriers to entry. The players *could* set up a cartel, price-fix at the monopoly price and share the revenues, but in most countries this is illegal, and in any case there is a temptation to cheat and lower prices to win market share. This can lead to ruinous price wars that bankrupt the weaker players. Oligopolists prefer to implicitly collude on price and differentiate on other attributes such as service quality or product characteristics. If one of the players is significantly larger than the others, it can punish weaker players by predatory pricing or using dirty tricks. As a result, the other players are often content to let the dominant player set the market price, and then to adopt their own non-threatening price and services strategies around it, simulating a competitive market. Because of the appearance of competition, regulators are often less aggressive towards oligopolies than they are to a clear monopoly supplier. The customer usually experiences an oligopolistic market as one with high prices and intense sales and marketing activity on secondary issues (e.g., complex price plans).

Telecoms suffer from two problems that make the pressures to bureaucratization worse: *service stability* and *process-intensiveness*. These two problems unfortunately potentiate each other.

Service Stability

Many industries have to innovate or die. Products have a limited lifespan and are then replaced by something quite different. Product turnover churns the organization and breaks up sedimented structures. But carriers are not like this. The chief product of carriers, and the one that still generates the bulk of their revenues, is the voice call—a product that has not changed substantially for the customer in more than 100 years, despite underlying technology changes. Other products have similar longevity. Once the transmission networks were digitized in the late 1960s, for example, it was possible to offer bit transport services. Apart from higher speeds and better reliability, these services have not changed fundamentally since then. Service stability means little outside pressure to change.

Process-Intensiveness

Process-intensiveness means that carriers depend upon a staggering number of routine and interlocked processes. Carriers are meant to be stable organizations, and with a conservative product set, they achieve this stability by a deep attachment to the processes they've got. Even discontinuous technology change such as the transition from analogue to digital switching, and from PDH to SDH transmission, left many processes unchanged.

This is true psychologically as well. Many carrier employees have a vocational sense of their jobs. Keeping the network going is a profound public service duty. When very senior managers present initiatives for change, often in very broad-brush terms, the reaction of many lower and middle managers is to interpret these as ill-informed, ignorant and sometimes malevolent initiatives whose only outcome will be to damage services. Their consequential defensive manoeuvres usually succeed. Senior executives talk of organizational immobility and sabotage.

How Carriers Responded to the Internet Challenge

The reader may be puzzled as to how such conservative organizations responded to the challenge of the Internet in 1998-2002. Apparently they all succeeded in becoming IP companies, didn't they? The truth is somewhat different.

Carriers were aware in the late 1990s that the Internet was becoming a new market, and many of them set up divisions to carry Internet traffic. These new groups installed large enterprise routers, all that were available at that time, which were usually connected into the carrier's ATM network as a provider of scalable bandwidth and inter-router virtual circuits for IP traffic engineering. The new Internet division did not impact any of the carrier's other divisions and processes, of course. The Internet division became the carrier's ISP, offering services such as dial-up internet access, e-mail, Web hosting and a portal. Staff often operated at arm's length from the regular divisions, dressed and ate differently, and worked unusual hours. They were tolerated by carrier management, and this toleration was met with disdain for "the suits" in return.

As the Internet became a dominant force in the world, its impact gradually became existential for the carriers. Around the year 2000, at senior executive level, it was finally realised that Internet technologies, products and services were the future. This meant that the hitherto standard response to the Internet challenge—treating Internet platforms as yet another overlay network with an incremental set of products—was not acceptable for the future. This was not Frame Relay all over again. Instead a more profound scenario presented itself: that on a foreseeable time scale, most of the current carrier networks and services would vanish and would be replaced by new IP based networks and services. For the existing networks and services, the Internet was not simply an "add," but a "migrate and remove" as well.

Options for Change

Senior executives faced two options for change:

- Create a new IP start-up with a new CEO, a distinctive management style, new processes, new IT systems and a network platform, and new staff. Gradually transfer customers across, eventually shutting down the current business.
- Set up an internal program to organically transition the organization from its current set of products, processes, and networks to the next-generation network, and *its* products and processes.

Based on the discussion on bureaucratization above, it might be concluded that this is a "no-brainer": only the first option, to create a new business, has a prayer of success. It is therefore interesting to discuss why almost every carrier nevertheless opted for the second approach, attempting to change organically.

Senior telecoms executives live in a world of projects and programs, often large scale. They are used to spending significant amounts of capital on innovation. Most of the time, these programs are additive, bringing new networks or products into being. Executives seem to have considered transition to the new IP base—now being called the "Next-Generation Network"—as a similar kind of project.

Executives at the VP and SVP level, owning major departments, had a career interest in participating in the creation of the NGN and could marshal some plausible arguments. The SVP for Product Management already had IP products and could argue that the process of migrating between "old-wave" traditional products and "new-wave" IP products (Table 5.1) could best be carried out within one department, his own, rather than between two different companies. Network Engineering and Operations could similarly point out that issues of transmission equipment, space and power in equipment rooms, and many network operation systems were common between the old and new waves, so there was no benefit in splitting them. In addition, the complex customer migration from the old platforms was best handled within one unified organization.

At the corporate level too, it was felt that close surveillance was required over both the present and future businesses. As the Internet boom faded in 2001–02, revenues were dropping away, CAPEX was becoming less available, and the timing of the transition more problematic. With so many interest groups in favor of a business-as-usual transition, and no real champion of change-via-spin-off, organic usually won. This partially explains why we have seen so little progress in transformation to date. However, if there *had* been such a champion, this is what he or she might have said.

"The SVPs have all made good arguments. Transition would indeed feel more comfortable and would be made administratively easier by being carried out

Table 5.1   "Old Wave" and Corresponding "New Wave" Products

Old Wave product	New Wave product
Internet Access over IP/MPLS	Internet Access over IP/MPLS
BGP/MPLS VPN	BGP/MPLS VPN
IP Security	IP Security
Voice	Voice over IP
Call Centre	Contact Centre
Frame Relay/ATM VPN	FR/ATM over MPLS Virtual Private LAN Service BGP/MPLS or IPsec VPN
Leased Lines	TDM traffic over MPLS
Fibre, wavelengths, Ethernet	Fibre, wavelengths, Ethernet

within an existing management structure. The people are used to working with each other and have good lines of interdepartmental communications. Notice that another way of describing what I have just said is: 'The recommendation is to execute the transition using our current processes for managing change.'

"But our current change-management processes are *completely incapable* of the fundamental *process re-engineering* and *IT systems innovation* that is a prerequisite for success. The level of investment needed to create a new competitive business with modern processes, Web-based customer and partner self-service, end-to-end flow-through automation, and IT application integration is just about imaginable. But the business case for such a root-and-branch transformation of the current, now legacy, organization, would never fly.

"Instead, the legacy organization should be locally optimized and run for cash. The new cannot fit into the interstices of the old. That is why the new IP services organization must be separate, must make its own decisions without having to integrate with the legacy of the old. If we don't do this, essentially nothing will happen."

It is not quite accurate to say that nothing subsequently happened. As 2001 progressed and the Internet bubble collapsed, there was a dramatic cut-back in market demand, as customers slashed their IT budgets, and ceased their previously frenzied acquisition of leased lines, Internet VPNs, and outsourced hosting platforms. Telecoms revenues fell away, and impelled by the need to cover operational costs and the enormous debt repayments on their capital expenditures for network build-out, companies engaged in fierce competition, forcing prices down to near short-run marginal cost levels, way below total costs. Bankruptcies followed.

The carrier stock response to cash flow problems is to cut costs, and many companies shed labour in brutal waves of redundancies. It was surprising to many people how many staff could exit a telecoms company with no apparent impact on services—in some cases they even improved. What was happening was that processes were being pruned of surplus staff by competent and informed lower to middle managers, who were getting blanket headcount targets and who used all their knowledge and creativity so as not to allow the company to fail. Clearing out the slack was everywhere a *local* optimization since capital to fund major process re-engineering or IT platform regeneration was in short supply. Consequently, it did not move the NGN transition forward significantly.

As the industry recovered in 2004-05, the problems of transition to next-generation networking came back onto the agenda again. The carrier currently making the running is British Telecom (BT). Its 21st Century Network program (21CN) is the most ambitious program of any carrier in the world at the time of writing. BT might appear to be attempting the transition in the organic fashion criticized above, but appearances can be deceiving.

BT has made an organizational separation between BT Wholesale, which runs its network platforms, and BT Retail and BT Global Services, which offer services to customers and which buy network resources from BT Wholesale across a regulated interface. The 21CN program is within BT Wholesale. This insulates the transformation program from the customer-facing process and system complexities in BT Retail and Global Services.

BT had already launched a major program to modernize its Operations and Business Support Systems (BSS/OSS), moving away from its existing complex and monolithic proprietary applications to the world of standardized COTS (Commercial Off-The-Shelf) applications integrated by middleware. This provides a necessary precondition and driver for effective process re-engineering.

Finally, BT has committed to the program very publicly at CEO level. Like publicly giving up smoking, such authoritative, senior and open commitment guarantees a powerful impetus for change.

It is too soon to judge 21CN, which will not complete its roll-out until almost 2010. The 21st Century Network program is described in more detail in chapter 13, which includes an interview with BT's chief architect.

Transformation: How Carriers Fail

I now recount three case studies, based on my own personal experience, where carriers tried to solve real transformational problems based on their in-house culture of expertise and process—and failed. The failures are instructive.

New Processes for a New Network

In the late nineties, I was technical architect of a major carrier network transformation program, part of which involved migrating from the earlier PDH transmission ("asynchronous" in North America) to SDH. PDH (Plesiochronous Digital Hierarchy—don't ask) and SDH (Synchronous Digital Hierarchy) are standards and technologies for carrying (multiplexing) many simultaneous voice and data connections onto high-speed links, usually fiber, and were discussed in chapter 2.

PDH was the earlier technology, usually deployed in a tree and branch configuration, with primitive instrumentation for monitoring and control. By the nineties, electronics and computing had advanced so that the new generation of more accurately timed SDH equipment could support automatic provisioning, failure detection and automatic recovery. Deployment was usually in rings: following a link or node failure, traffic was routed the other way round the ring. The new SDH network allowed far more automation of circuit management, both for circuits connecting the carrier's own voice and data switches and for leased-lines sold to customers, so clearly new processes were required.

I remember a meeting, where the urbane and civilized SVP in charge of Operations called his staff together to discuss how new processes could be designed and established to get the benefit of the state-of-the-art SDH network we were so expensively installing. There must have been at least 40 transmission middle managers in that large conference room. The SVP adopted a bottom-up facilitative management style, encouraging ideas, answers and solutions to come from the floor. Two hours later, after much opinionated turf-protecting and lack of engagement with the real issues (which hardly anyone at the meeting actually understood, anyway), a series of irrelevant action points were agreed for the next meeting. Subsequent meetings over a period of months predictably advanced matters not at all.

The change management procedure that actually solved the problem was to establish a separate group from scratch to take customer orders and provision SDH circuits. Roles allocated within this small group were designed to exploit the power of the new automation systems, creating a streamlined division of labour not possible with the existing primitive and mostly manual PDH processes.

Management Lesson

When faced with discontinuous innovation, a bottom-up consensus-driven procedure based on the existing accountability holders, with little preparation or top-level direction, does not work.

Strategic Transformation Towards an IP Services Company

I was also, a little later in my career, privileged to watch the opposite management style. Detailed directives for change from a hard-driving visionary senior executive, given as specific action points to his subordinates. But his attention immediately returned to high-strategy: he was completely uninterested in the daily hard grind of program execution and the need for continual oversight of problems and progress. Of course, none of his hapless subordinates were able to force change through their own sub-bureaucracies. Only a combination of ingenious excuses and the opportune arrival of numerous attention-diverting crises kept them safe. Finally the visionary executive was forced out by the greater failures of his organization.

It has been popular in recent years to identify *execution* as the single most important factor in business success. The basis of execution, however, is to understand *what has to be done to prevent the existing bureaucracy subverting necessary change*. This is the *hardest* problem, but the visionary thought that exhortation would do it, and that the details could be delegated. This is an all-too-common response.

Management Lesson

Management by vision and will-power alone is not enough. Supervision of execution is vital.

IT Outsourcing

One of my consultancy assignments was with a carrier that had outsourced its IT to a large systems organization I will call "Bravura." Under the new regime, Bravura was responsible for desktop support, server management, the IT roadmap, and transitioning the IT infrastructure towards the state of the art. In exchange, the carrier paid Bravura extremely large sums of money both on a rental basis, and for jobs done. When the deal had been agreed, most of the carrier's IT staff had moved across to Bravura as employees.

Why had the carrier done this? Around the time of the deal, it had been hard to miss the deep smiles of satisfaction on the faces of the COO and his colleagues. They had wrestled for years with intractable problems of IT service quality and modernization. The only results had been regular service outages ("the e-mail is down again") and appalling service from what was euphemistically-termed the "help desk." The truth was, IT was out of control and unfixable, and now they had made it Bravura's problem.

The first change the carrier noted had nothing to do with IT service quality, it was the quality of the Bravura staff they were dealing with. While the deal was in play (and it was financially enormous) the COO had been dealing with Bravura's A-team. With the contract done-and-dusted, it seemed that the A-team had gone off to fight for other deals in other parts of the world, and that Bravura's post-sales B-team had arrived to actually do the outsource. It did not feel good.

Pre-contract, if you had an IT problem, for example ordering new PCs, fixing a desktop fault, or installing new applications, you would get onto the help desk, get a ticket number, and wait ... and wait. Under the new regime, the waiting was relatively unaffected, but you now had to fill in forms. Lots of them. Since Bravura charged for work done, everything had to be documented in detail, and then signed off both by carrier staff and by Bravura managers. IT-related activities slowed to a crawl while user dissatisfaction reached new highs.

The IT transformation part of the contract was likewise not in the healthiest state. Bravura had assigned a chief architect and team to its outsource organization. They had spent quite a bit of time, and of course money, preparing an IT roadmap and transition plan. I know, because I had the privilege of looking at it. But few of the carrier's staff seemed that interested: the document was merely motherhood and apple-pie. Bravura's architects had no access to senior business thinking within the carrier and had little to no idea of its business, product, and network strategy. They also found it difficult to inventorize the enormously complex set of legacy applications and interfaces, many of which were undocumented and some of which ran on bootleg servers under people's desks. It was unsurprising that they had served up a bland, standard industry roadmap with no touch points to the carrier's IT reality at all.

After a few years of poor service, huge bills, IT immobility, and endless legal wrangling the carrier managed to terminate its contract with Bravura and brought the whole mess in-house again. Where it remains.

Management Lesson

- Don't outsource a mess.
- IT is a core competence for a carrier. Only outsource those functions that are clearly commodities.

On Doing It Right

My conclusions over the years have remained unchanged. Successful change management requires:

- A clear idea of what the objective actually is,
- Support from the most senior executives,
- A program team with process re-engineering, architecture and design, and program management expertise,
- The program team must be full-time, and distinct from the line organization,
- The program team needs authorized touch points at all levels with the line organization.

And, most importantly, find exceptional people to fill the program team roles, critically the senior positions. These points are explored more fully in chapter 7.

Reorganization

It is traditional in discussing the problem of organizational change to talk about reorganization, and to display the apt quote below, fallaciously attributed to the

Roman satirist Petronius Arbiter, but apparently penned by a literate British soldier in Germany after the Second World War (Sullivan 1981).

> We trained hard, but it seemed that every time we were beginning to form up into teams, we would be reorganized. I was to learn later in life that we tend to meet any new situation by reorganizing; and a wonderful method it can be for creating the illusion of progress while producing confusion, inefficiency, and demoralization.

In fact I have found reorganizations, both useful and pointless, to be more of a feature of vendor organizations than carriers. Nortel used to cycle its global business structure every three to four years. Take an arbitrary point in the cycle where power was with the country and regional sales and marketing organizations. They had control of spending, and accountabilities for profit and loss, and they placed orders on the back-office product divisions. Over time, they would grow fat and comfortable, recruiting endless sales support staff and constructing lucrative private empires. SG&A costs would soar.

After a few years of this, Nortel would abolish the structure and would give control of marketing and sales to the product divisions. The switch guys would sell globally, as would the transmission guys, the data guys, and the wireless guys. Marketing was separated out with only a thin overlay of corporate strategic marketing. Account teams were in common, but were normally owned by the lead product division for that customer.

This cyclical reorganization was always intensely controversial internally, but had the effect of smashing up empires and bureaucracies, and reducing overmanning. After three or four years of watching sales and marketing duplication emerge in the Line of Business model, with product division SG&A remorselessly increasing, Nortel would switch back to its original model, and lose another set of people and costs.

Carriers don't reorganize enough. The stability of their markets and products, already alluded to, seem to give them little incentive to change how they do their businesses. Individuals often get switched into and out of jobs, but the underlying structures and processes are unaffected—senior executives too often are "'in office, but not in power."

For most of my career dealing with carriers, I was strongly of the opinion that that the answer was to break the carrier up into smaller, market-facing units with P&L responsibilities, where competitive pressures would counteract inefficiency and bureaucracy. Unfortunately the laws of economics are against such a strategy. Telecoms is all about economies of scale, and a naive cottage industry approach doesn't really work. But perhaps something similar does.

One of the most encouraging developments of the last few years has been the rise of the virtual operators. With the process-centric factory functions of running networks off their hands, they can organize around the principles of customer service, and the results are often exceptional. As the provision of network services continues to commoditise, an industry model that more sharply delineates between wholesalers and retailers can result in more efficient and competitive companies that each serve their specific customer types better. This is a future worth fighting for.

Summary

In this chapter we have looked at the problems of carrier bureaucratization and resistance to change. This is partially due to the concentrated nature of the industry and partially due to the nature of the telecoms business itself. We then looked at the challenge of the Internet, and how two responses were possible: either to spin-off a new IP company, or to attempt an organic transition to the next-generation network. We discussed why almost every carrier decided upon the latter course.

We then looked at three failure scenarios: SDH process innovation, strategic redirection, and IT outsourcing, and drew some lessons, and finally we looked at reorganization—often considered as wholly negative—and considered whether the problem might be that carriers don't in fact reorganize enough. Finally, it was suggested that the wholesale-retail split we already see with virtual operators may promise considerably improved operation and greater flexibility in the future.

Appendix 5.1: What Do Carriers Actually Do?

A friend of mine once told me that telecom operators are secretly delighted when the staff goes on strike. Only about one job in ten in a carrier is actually dedicated to providing day-to-day operational service to customers. Operation is, in any case, pretty automated and middle managers, most of whom were promoted through the ranks, can usually step into any gaps. The operator saves a fortune in wage bills.

So what are the other nine out of ten people doing? Planning for the future, mostly, manning the functions shown in Figure 5.1.

Operational Processes

This is what the customer—residential, corporate, or fellow operator—sees. The carrier markets and sells its products, takes orders, provisions the service onto its network, operates the service, fixes faults, and bills the customer. Call centers for customer care are included here.

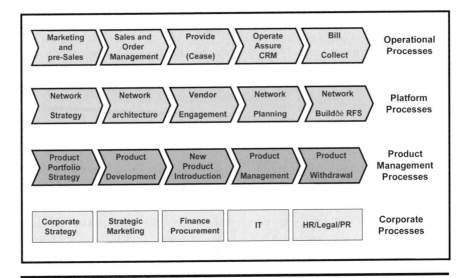

Figure 5.1   Telecoms Operator Processes.

Platform Processes

The carrier's main asset is its network. As technology develops, there is a continual process of:

- Identifying new technology opportunities (often suggested by vendors),
- Developing a network and systems strategy,
- Designing an architecture appropriate for the carrier's network,
- Engaging with the vendors through RFIs, RFPs, RFQs (see below),
- Running a procurement process for vendor selection (this is an auction—see chapter 10);
- Finally planning the new platform or network in detail, and installing it.

Many of these functions can be carried out in partnership with a preferred vendor, or even outsourced to it through a longer-term frame agreement.

Product Management Processes

Products go through a lifecycle like anything else, and not all products are successful. There has to be a way of identifying new product possibilities, and of managing new products into service. Similarly, when a product's costs outweigh its revenues as it comes to the end of its life, there has to be a way of retiring the product and managing its remaining customers off the service, often by putting the price up prior to simply withdrawing the service.

Corporate Processes

These are the standard support functions: corporate strategy, strategic marketing, finance, procurement, HR, legal, PR, and IT. They are often organized centrally across the group.

Standard Models of Telecoms Processes

Constructing carrier process models is not easy—it amounts to defining a theory of carrier organization. It is so difficult in fact that specialized consortia have come together to define best practice. Organizations such as the TeleManagement Forum (http://www.tmforum.org/) have developed complex models such as eTOM, the enhanced Telecom Operations Map®, to standardize carrier process models via layers of process abstraction.

In my experience, these are more talked about than actually used by executives who take charge of reorganizations.

Appendix 5.2: RFIs, RFPs, and RFQs

Before I first entered the telecoms industry, I had been doing computer science research for most of the previous decade. I suddenly had to learn a whole new set of acronyms and processes. One area that was completely new to me was the intricate choreography between vendor and customer—the dance of the sales process.

The process of engagement between a carrier and a vendor often starts with the carrier perceiving it has a problem. Perhaps it ought to do something about video over IP, for example. If it genuinely wants to better understand the potential opportunity, it can solicit some free consultancy from potential vendors, by issuing an RFI (Request for Information). This document describes the carrier's perception of the problem or opportunity, and asks for information from the vendor as to how the vendor suggests the carrier might address it. The account team within the vendor organization put together a task force and staff it with the appropriate expertise under a project manager to put the document together. Quite often a vendor has developed a new technology and thinks it would be interesting to the carrier, so makes an unsolicited bid. So carrier-pull is blended with vendor-push.

Once the carrier has a clearer idea of what it ought to want, the next stage is to assess the capabilities of different vendors' equipment. An RFP (Request for Product) is issued, and the carrier can compare the responses from different vendors.

As we get nearer an implementation project, the carrier invites vendors to bid for the business through an RFQ (Request for Quote). The response will describe the vendor's proposed solution, and include a commercial section detailing pricing. The collection of received RFQs, together with vendor presentations, serve to narrow the field down to a final short list. Then real commercial pressure is applied to negotiate prices down still further.

It would be quite unusual to see a closely coupled sequence of RFI-RFP-RFQ as part of a defined sales sequence because of the overheads involved and the overlap in content between the deliverables. What tends to happen in practice is that some part of the carrier organization, often low in the hierarchy, is alerted to a new opportunity and issues an RFI to create a framework for discussion with their vendors. After the RFI process has completed, the issue then diffuses within the carrier and either dies, or accumulates momentum so that a tentative decision is made to do something in this area. An RFP or RFQ process is then launched in parallel with a business case being put together. For larger projects, a formal RFQ process is the mechanism of choice for vendor selection. In the last lap, it's all about cutting deals and sometimes it's not very pretty.

References

Sullivan, J. P. 1981. *Petronian Society Newsletter,* Vol. 12, No. 1: 5, May 1981. http://www.chss.montclair.edu/classics/petron/PSN1112.5.GIF. A discussion of the erroneous ascription can be found at: http://www.dtc.umn.edu/~reedsj/petronius.html.

Chapter 6

Telecoms Market Structure

The Rule of Three

In 2002, two management researchers, Sheth and Sisodia published a book called *The Rule of Three*. In it, they analyzed pretty much every market they could get data on (144 market sectors across the world) and found remarkable similarities. It turns out that in a naturally competitive and mature market, without excessive regulation, the market segregates into three broad domains (Figure 6.1; Sheth and Sisodia 2002, 4).

The first domain is defined by *three* major players, generalists who together control around 70–90 percent of the market. The second domain comprises niche specialists, each holding a monopolistic position in a tightly-defined product or sub-market category—each specialist typically has a market share of 1–5 percent, but makes high margins due to its niche-market dominance. The third domain is termed "the ditch." It consists of those companies with around 5–10 percent of the market who lack the scale economies of the first group, or the niche focus of the second. The ditch can lead to bankruptcy or takeover. Consider the example of the shopping mall, anchored by the major department stores, the generalists, but with many niche shops along the mall corridors connecting them, with some of them clearly struggling.

In economic textbooks we are told that more competition is always better for the consumer. Price is beaten down to long-term marginal cost, and all social welfare benefits go to the consumer. However, this is an idealized, steady-state model—it permits the producing company little margin for slack. In markets subject to uncertainty, discontinuities, and continual innovation, companies need to bet on risky decisions; this requires capital reserves that are often unattainable in conditions of extreme competition. Massive competition can also hurt the customer, as we see today in many consumer electronics markets where buying

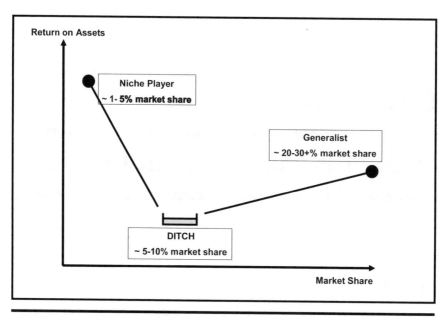

Figure 6.1 Market Structure.

decisions are confused by excessive choice, and not every vendor will make enough sales to survive. The situation of hypercompetition, typical of early markets or markets emerging from a major market discontinuity, leads to shakeouts followed by the market structure defined by the "Rule of Three," large-scale oligopoly plus small scale monopolistic competition.

Considering first the large oligopolistic generalists—why should there just be three? The authors argue that duopoly is unstable: the two players either attack each other destructively or collude, attracting regulation. With three players, any two can cooperate against the third, maintaining a balance of power.

So why not four? The authors speculate that consumers value a manageable choice between three suppliers, but that further choice just creates clutter, confusing the market. However, the dilution of market share with four major players can also lead to instabilities, driving the weakest into the ditch. The number one player should also be careful about gaining too large a market share, the authors argue: past 40 percent, regulative scrutiny becomes more intense, the proportion of underperforming customers increases and growth becomes harder.

The Rule of Three therefore represents a compromise between sufficient competition and sufficient market share, but it can be distorted by factors such as regulated monopoly, major barriers to trade (e.g., in global markets), a high degree of vertical integration impeding consolidation, and a history of monopoly prior to deregulation. All these factors are clearly prevalent in the telecoms sector.

Sheth and Sisodia (2002) had advice for each of the generalists in the major, oligopolistic sector of the market.

If you are number one in your market, you should:

- Be a fast-follower—learn from the number three company, which tends to be more innovative;
- Make your standards those of the industry (a well-known Microsoft tactic);
- Exploit your position with strong branding (familiar from Cisco, Intel, and many others);
- Grow market and focus on volume, not margin (exploit economies of scale).

If you are number two you should:

- Combat the market leader with better, more focused marketing;
- Cherry-pick the best customers;
- Compete on price and focus on value-added services;
- Either challenge the leader, or segment the market.

Finally, if you are number three, you should:

- Outflank numbers one and two by product and process innovation;
- Partner with suppliers or customers to defend your market share;
- Grow by acquisitions from "the ditch."

Conversely, if you are a specialist, in the 1–5 percent niche part of the market, you should:

- Stay resolutely focused on the niche—do not be tempted to overdiversify and lose touch with your core market;
- Shun fixed costs—they can detract from flexibility;
- Create entry barriers, e.g., by negotiating sole rights with suppliers, by establishing local monopolies.

General features observed across many markets include:

- If a market leader has approximately 70 percent of the market, there is no room for a second generalist, but the situation is unstable;
- If the market leader has 50–70 percent of the market, there is no room for a third generalist, but one will eventually reappear;
- If a market leader has less than 40 percent of the market, there may be temporary room for a fourth generalist, but the ditch beckons.
- In a downturn, the battle between generalists one and two can send number three into the ditch while specialists remain unaffected.

Application to Telecoms

It is easy to see the relevance of this analysis to the North American fixed telecoms market. Let's take a look at the historic long distance carrier structure of AT&T, MCI, and Sprint. Recent acquisitions have led to AT&T taking the number one slot with Verizon as number two. Given the power and market share of these two, there is scant room for a third player, and the ditch looms for Qwest and Sprint, as they seek a more profitable destiny through more creative and perhaps diversified business models.

In Europe, the continuing dominance of national markets, the consequences of deregulation in the 1990s followed by the Internet-boom and investment in new facilities-based carriers and a history of monopoly has resulted in a fragmented market structure with too many small players. We are still in a hypercompetitive phase with shakeouts and further mergers and acquisitions to come. France Telecom (using Orange as its flagship brand), Telefonica, and Deutsche Telekom have all used their financial muscle to aggressively expand within Europe, and to acquire both fixed and mobile operators. At time of writing, Vodafone and BT, both British-owned, stand out as respectively mobile and fixed pure-plays. Neither company's strategy seems stable for the longer term.

The UK is generally considered the most competitive market in Europe, due to the impact of early deregulation, yet its immaturity is shown in Table 6.1 (revenue shares of a representative set of operators based on 2002–03 figures, but there have not been substantial shifts as the market has been flat).

It should be noted that these are aggregated figures, taken from annual reports. In particular markets, such as Broadband, BT will point out that the situation can be far more competitive. For example, BT's retail broadband product has been taken up by only a quarter of broadband subscribers, with Virgin (NTL-Telewest), Wanadoo (France Telecom), AOL, and Tiscali all strong competitors. The entry of Sky as a major broadband player will transform the market further.

Table 6.1 Operator Share of Revenues in the Early 2000s

UK Operator	Percent Revenues
BT	78.0%
C&W-Energis	10.4%
NTL-Telewest	7.1%
Verizon UK	1.8%
COLT	1.4%
THUS	1.2%
Fibernet	0.1%
Total	**100%**

In the mobile sector, according to 2005 reports from the UK regulator Ofcom, the four major mobile network operators Vodafone, O2 (Telefonica), T-Mobile (Deutsche Telekom), and Orange (France Telecom) had roughly equal market penetration. Due to its economies of scale, and strong corporate presence, Vodafone is the market leader with around 30 percent of the revenue pie, with T-Mobile dragging at 19 percent. The 3G operator three, relatively new to the scene, has grown its revenues to around 5 percent.

The Rule of Three is violated here, with one of the reasons being the deep pockets of the owners of Orange, O2, and T-Mobile. Arguably the true market for these mobile operators, along with Vodafone, is Europe itself, rather than any one national market.

The Consequences of Market Immaturity

There are too many facilities-based carriers. The UK alone has at least 12 completely functional national fiber networks, mostly built in the 1990s deregulation and during the Internet bubble of 1998–2002. A partial list would include:

- BT,
- Cable & Wireless,
- NTL/Telewest,
- Energis (National Electricity Grid network),
- Kingston Communications (Torch network),
- Thus (Scottish Power),
- Geo (Hutchison, 186k, Transco Gas network),
- Sky/Easynet (British Waterways network),
- Verizon UK,
- Viatel (AT&T network),
- Global Crossing (Racal, British Rail network),
- Fibernet.

Many of these are part of a broader pan-European network.

The inevitable consequence is that prices for many commodity network services, such as leased lines and Internet access, are closer to short-term marginal cost-to-provide than long-term incremental cost (quite a hit higher, as it has to provide for the next round of capital spending and cover broader overheads).

With pricing set so low, the UK facilities-based market, for example, has split into three components.

- The former monopolist, BT, with economies of scale, a large inherited customer base, and stable revenue streams, especially from its consumer business, continues to be profitable.

■ Niche players, with lower operating costs and modern IP-based networks, which hang on in expectation of a rising market.
■ The ditch, which now encompasses the other traditional carriers, such as Virgin-NTL-Telewest, and C&W-Energis. Here, a higher cost base due to diseconomies of scale results in less profits, and less capital reserves to effect the transition to next-generation networks. It is no coincidence that this is where we have seen recent mergers in a continuing M&A process, and much agonising about future business models.

Next-Generation Network Business Positioning

It is possible to put together a map showing the next generation network by layer, the new services each layer supports, and the delivery mechanism to a segmented market (Figure 6.2). If we start at the top, the standard telecoms market segmentation distinguishes consumers from business customers. The business sector is then, in its turn, further segmented into global enterprises, large national corporates and small to medium enterprises (SMEs). Sometime "small" is distinguished from "medium" in this last category.

Getting to consumers is always the problem with the residential market segment. Copper wire is normally owned by the incumbent. If cable has been dug, this is owned by the cable operator. Satellite works just fine, but is affordable only one-way. Wireless local loop has been tried many times, but has failed equally

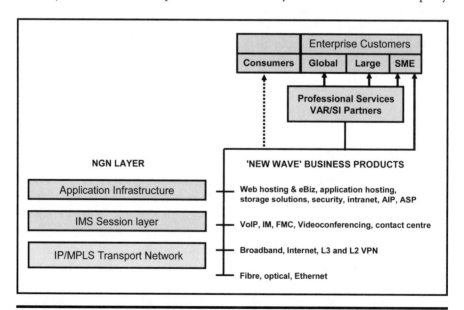

Figure 6.2   NGN layers and their "new wave" services.

often—perhaps municipal WiFi and WiMAX will eventually break the spell but it is hard to get costs down. People *will* pay a premium for wireless access, but they are paying for ubiquitous *mobility*, not services delivered specifically to the home location (reception is frequently poor in the home).

Regulation has, in some cases, made the access "last mile" link a level playing field for any operator by controlling wholesale pricing and exchange-access. Broadband via local loop unbundling is a case in point. But for many alternate operators, the opportunity for premium-return communications services to the residential sector are just too meager, and they have abandoned this segment, or consigned it to a niche part of their businesses. (Content providers such as cable and satellite companies are in no such predicament—see chapter 15.)

At the other extreme, global enterprises, as we will discuss in the final part of the book, are hard to service unless the carrier is also a global player, and profit margins can be very thin. This means that most alternative operators are focused on large corporates and SMEs.

Strategies for Telecoms Generalists

A generalist already has scale and a large market share, typically in excess of 20–25 percent. This will support a full portfolio of products addressed to many market segments. In fact, incumbent carriers across the world tend to offer a similar portfolio, with a capability to colonies new opportunities as they arise. Quadruple play is particularly attractive, as it leverages the full revenue possibilities of high-speed broadband access (for voice, data, and TV) and combines it with mobility. Across the world we see incumbent carriers, major cable companies, and satellite-based broadcasters contesting this space.

Strategies for Telecoms Niche Players

The broad sweep of the generalists creates major problems for smaller alternate network operators—can they compete? As Sheth and Sisodia recommend, they have to specialize. Their choices would appear to be either by product or by market sector, but this is illusory: horizontal telecoms products achieve their ubiquity through their commodity status. Not a place for premium returns unless the cost base is beneath the floor. So, although there are a few companies out there that will sell undifferentiated, POTS, vanilla broadband, dark fiber, or colo space to anyone at all, most specialists seek to service a specific market ecosystem: the sweet spots are corporates, or SMEs.

The next-generation network enables a combination of connectivity, communications and application services to be provided for the corporate and SME sectors. Connectivity services include various kinds of VPN and leased lines, while communications services include voice over IP, e-mail, instant messaging, video-conferencing, and contact centers. Application services include application and Web hosting, data storage and backup, and value-added services such as security.

Note, again from Figure 6.2, the important role of value-added resellers (VARs) and systems integrators (SIs) in delivering these services to customers. It is rarely the case that they can be bought like groceries from a supermarket. Essentially all of these services require careful configuration to suit the business's needs, and integration with their existing networking and application infrastructure. This takes skilled people, either a professional services arm of the carrier, or a partner organisation specializing in systems work. VARs and SIs exist on all scales from global companies such as IBM, EDS and Accenture right down to small companies operating within just one town.

From the point of view of a VAR or SI, their job is to analyze the customer's requirements, to design a solution, to buy in commodity service components from carriers and telecoms/IT vendors, and then to integrate everything together as specified in the solution design. There is a considerable overlap in this mission with the business objectives of the carrier (Figure 6.3).

Carriers have wanted to "move up the value chain" into professional services for a long time. Success has, however, generally eluded them. Professional services is a people and skills business, while telecoms is a process and routine business. Professional services tends to generate revenues based on idiosyncratic projects done, while telecoms generates recurring revenues on utility-delivered services. It is the recurring tension between hunting and farming, to use the sales metaphor.

Clients tend to prefer dealing with independent VARs. They welcome the em-

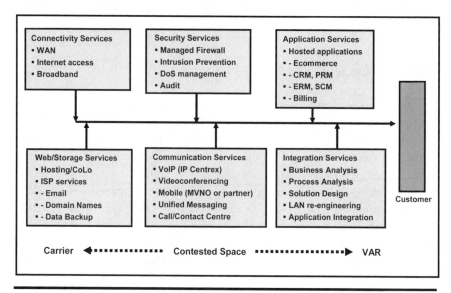

Figure 6.3   Near-future services and the value network.

phasis on personal service and the independence from any one telecom supplier. Businesses have been burned too many times by carrier lock-in.

From a carrier point of view, despite their aspirations to value-added services, they believe in their hearts that most of their revenues will come from telecoms services annuities. They forbid their professional services arm from shopping around to their competitors, and throw their SI services in as a cut-price sweetener, or even for free, to clinch a deal. The professional services division is perpetually confronted by the dilemma: am I a profit center, or am I a customer operations cost center? They would like to be the former, but those carrier instincts keep pushing them to the latter position.

The imminent arrival of the next-generation network *investment step-function* accentuates this dilemma to an extreme degree. Insofar as alternate operators cannot afford to invest in full-scale next-generation networks of their own, they have the alternative of buying NGN services from bigger players on the wholesale market. But in doing so, they will be operating more like VARs and SIs than carriers. The trajectory from carrier-centricity through to VAR-centricity may be the most intelligent response to avoid being dumped in the ditch, but timing is crucial. The large incumbent generalists will not necessarily wish to offer key NGN interfaces in the wholesale market to protect their own investment in "new wave" services, and regulation may be light for a whole period to protect that investment.

The best response might be to selectively invest in those aspects of the NGN that underpin the service portfolio being offered to the niche market. At time of writing, Kingston Communications in the UK is an example of a carrier that has acquired medium-scale VARs into its Affiniti division, aimed at corporates and SMEs. Affiniti has taken the view of a Systems Integrator and is prepared to buy telecoms services from a variety of suppliers, not just Kingston itself. It is likely, though, that they will see continuing advantages to retaining their own network, and in acquiring targeted next-generation network functionality as their clients come to need it, and insofar as the wholesale market is not providing it.

A further advantage of targeted, just-in-time investment in NGN components is that costs are likely to be lower. Most NGN technology is software, which is market-priced by the vendors (marginal cost being close to zero). The *incumbents* are forcing through their NGN programs for strategic reasons, and in the process paying some/most of the vendors' development bills. But a small alternate operator should be able to bid prices down a few years out, at a point when NGN "new wave" services finally become real. Chapter 14 covers alternate operator strategies in more detail.

The NGN Ditch

The next-generation network ditch will be defined by those carriers that invest in an NGN without a clear focus of what it will be used for. It is worth emphasizing that the "out-of-the-box" NGN is really a piece of middleware—generic IP transport, session management, and application hosting functionality—in need of significant further investment to add specific products and services. A sub-scale player is going to have problems all the way along, and is likely to find they are hemorrhaging cash en route.

For example, if we take the UK situation at time of writing, we have two main ditch players, Cable & Wireless UK (recently merged with Energis and Bulldog Broadband) and the NTL-Telewest-Virgin merger. The UK market is dominated by BT, and its 21CN evolution to a transformed NGN-based organisation, and the dark horse is Sky, with its potential to leverage the IP alt-net acquisition of Easynet.

Cable & Wireless UK has already retreated from almost all of its traditional market segments to concentrate on UK corporates (where it is head-to-head with BT Retail and Global Services). Its central problems of legacy and lack of scale continue to haunt it, and its eventual fate seems to be acquisition by someone with enough capital to fix it—a three to five-year project. The present management seems to be clearing the decks to this end.

The NTL-Telewest-Virgin merged company has considerably more promise. It has an extensive cable access network in large parts of the UK that opens up numerous broadband possibilities. It has a consumer content business (cable TV) and, with the Virgin acquisition, it is now an MVNO. It is starting very much in the Sky/Easynet space but with more of a carrier culture, which could allow it to move faster into business services. It is well-placed for the number two slot behind BT as the second generalist, and taking Sheth and Sisodia's advice for a number two player, it should choose its battle carefully.

Summary

In this chapter we have reviewed Sheth and Sisodia's concepts of market structure and applied them to the telecoms sector. The transition to next-generation networking, with its major capital requirements and dislocation to existing business models and processes, exposes the strengths and weaknesses of all the players. The future generalists need to understand their future portfolios, the future specialists need to understand what business they will really be in, not necessarily as carrier-centric as they have been in the past, and the ditch-dwellers need to decide if they can become future generalists, or if perhaps they don't have a future at all.

Many of the topics flagged in this chapter are discussed in much more detail in the fourth part of this book, which deals with business strategies.

References

Sheth, J. and Sisodia, R. 2002. *The rule of three*, 1st ed. New York: Free Press.

Chapter 7

Choosing the Right People

Introduction

The intention of this book is to prepare its readers for the challenges of the Next-Generation Network in the broadest context. Some of these challenges are technical. It is quite important to understand what the NGN actually is; even if there are specialists "out there" who understand the various layers in the greatest detail. Other challenges relate to business models and marketing strategies, topics discussed in preceding chapters, and to which we return in the final part of the book. But people are also part of the NGN, not only with regard to identifying the right roles in the future organization, and finding the right people to inhabit them, but also with regard to understanding how to define the roles and select the people to accomplish the NGN *transformation* itself.

Managers should be surprised at how much is now known about classifying personality and intelligence. An enormous amount of research has been carried out into the attributes required to successfully accomplish various roles—research that has yielded a high degree of consensus. Yet many decision makers are completely unaware of the frameworks and tools at their disposal, preferring to subcontract to HR or to consultants. This chapter should serve as a wake-up call, to give you, the reader, a background in understanding personal differences, helping to deal effectively with different kinds of people, and choosing the right individuals for the right tasks. It may even help your own career, and perhaps provide you with greater insight into how you are likely to be perceived by other people, in work and out of it.

Individual Personality and Intelligence Matters

How many times have we railed against senior management when they allow a talented person to leave or be transferred, forcing the appointment of a mediocre person to an important position. People are not interchangeable. People are not components to be picked up and slotted into place.

The monopoly carrier of the past was indeed, on the face of it, populated by an army of faceless clerks and technicians. But even the modest deregulation of the 1990s was sufficient to drive competition, forcing the removal of layers of these employees, and encouraging the recruitment of bright people with flair. New disciplines (at least to carriers) such as marketing were finally recognized as important, and were staffed with talent. The better vendors had always known this.

When Nortel decided that the Internet was real, in the late 1990s, it undertook a "right-angled turn." Managers were put through a three-day personal reevaluation workshop. We were sent to a pleasant hotel by the seaside (out of season, though) and given batteries of psychometric tests. We were conducted through videotaped role-playing and were counseled on career options. The extremely competent consultants and the relatively small team-size allowed a very personal style of guidance and the result was perhaps the most impressive workshop I have ever attended. Many people, myself included, left with new thoughts about ourselves, our aspirations and what we ought to be doing to actively develop our careers. That was precisely the intention, of course.

Nortel had to cancel this exercise prematurely—the cost was exorbitant—but the basic premise was spot-on. People *are* a factor in the corporate success vs. failure equation, and there exist perfectly good tools that can be used both to classify and develop people as regards to their career options, and to steer them towards the most appropriate roles.

Pre-Nortel, I had been used to Dilbert-style managers: concrete, administratively-minded individuals obsessed with the minutiae of cost-control and routine. At Nortel, I found completely different management styles.

The leader of my division was Alicia. Highly-intelligent, Alicia was also quiet, self-effacing, and almost completely nondirective in her management style. She would chair meetings in an almost feline fashion, asking everyone to contribute, making sure that no one was left out, and letting consensus emerge from the collective discussion. I found the style both transformationally refreshing and completely bewildering: here was a leader who, apparently, did not lead. I had not yet heard of the "servant leader" concept.

Alicia was outstandingly successful as the leader of a Professional Services consultancy group. On the strength of it, she was promoted to lead a development team. Here her style was less successful. Finally, Nortel seemed at a loss to know what to do with her, and she entered that shadowy world where you do senior odd-jobs: speaking at conferences, representing the company at B-list events, the ante-room of the way out. Alicia is now successful, but not with Nortel.

Mark was the opposite. Brought in to fast-track the introduction of a new carrier product, Mark was a maverick who broke the rules. He was smart and completely task focused. Based in Ottawa, he would think nothing of calling people at home in the evening (his afternoon) and instructing them to produce something there and then. Weekends and holidays meant nothing to him, particularly as he had staff he could delegate to. We grew to dread his calls.

Mark achieved his mission and was duly decorated by senior management. He moved on to oppress another bunch of staff to the relief of those of us for whom the nightmare had just ended. In the end, he made just too many enemies and is currently running businesses a long way from Europe.

Piers, by contrast, cared about his people. He knew most of his large staff by name, and did not forget birthdays. His meetings were long and detailed, and covered many, many issues. He had plans for reform, but they were hazy, and somehow the big changes never came. Despite a high workload and attention to detail, there was a larger-scale sense of drift and that classic feeling of rearrangement of deckchairs on the *Titanic*.

We recognize all these people, and more besides. They were all talented individuals who had come up through the Darwinian process of management selection. Were they the right people for those particular jobs? We need a language to discuss this in a standardized and objective way.

Types of People

When we were living in Virginia, I once attended Catholic Mass with my wife. The priest, addressing the large congregation, introduced the featured speaker, a Jesuit priest, "We are fortunate today to have Father Michael Smith SJ to talk to us. Sounds a bit like Myers-Briggs doesn't it?" (laughter). Fairfax County's professional congregation knew all about Myers-Briggs, because it was the personality assessment they had all taken at work. The Myers-Briggs Type Indicator™ (MBTI) is the most popular psychometric test in the world.

Isabel Myers and her mother, Katherine Cook Briggs, developed their theories during the Second World War. Their initial self-assessment test was based on concepts of psychological type developed originally by the Swiss psychologist Carl Jung, but Myers refined and extended the concepts, orienting the approach towards "normal people" rather than the clinically-referred individuals whom Jung had theorized about. The Myers-Briggs approach was developed outside the academic community, at that time mired in the morass of behaviorism. This separation continued, even as Myers-Briggs took the global corporate community by storm. More about this in Appendix 7.1.

The Myers-Briggs approach to personality differences is deceptively easy to state. Individuals taking the self-assessment inventory answer questions that produce scores on four dimensions as follows.

1. *Extraverted* or *Introverted?*

Is "hell at a party" not being able to get in? Or being there? Extraverts take their energy from socializing with other people and feel underenergized when alone. Introverts need time alone, and socializing makes them weary. Most people recognize themselves on one side or the other of this divide. This is the **E-I** scale.

2. *Sense-impressions* or *iNtuition?*

Some people deal with the world in a concrete pragmatic way, others grasp situations through a framework of concepts, values, or ideals. In the former category are many sports people, administrators, policemen; in the later category intellectuals, conceptual artists, campaigners for a cause. This is the **S-N** scale.

3. *Thinking* or *Feeling?*

Another poor choice of words from the founders. We all know people who are coldly focused on the logic of the situation (e.g., Mark, above), who will do what is logical, and take the interpersonal consequences. This applies as much to the logically-focused intellectual as to the results-focused executive and mission-oriented special forces soldier. On the other hand, there are those, like Piers (above), who put personal relationships first, who are motivated by sustaining the cohesion of the group, making their opponent a friend, or who are driven by a moral imperative to care.

The former score highly on the "'Thinking" side—sometimes the **T** is read as "tough-minded"; the latter score highly on the Feeling side, although feeling is not so much raw emotion as a genuine warmth, empathy and an orientation to human values and solidarity. This is the **T-F** scale. And by the way, men's scores tend to be skewed to **T** and women's to **F**, although as a corrective, we have all met women who are probably from Mars, and men who have real warmth and are natural hosts, entertainers, or diplomats.

4. *Judging* or *Perceiving*

This dimension is the staple of so many comedies, as well as real life dilemmas. One half of the partnership likes everything planned and organized in advance, the other hates lists, loves freedom and just wants to live life as it comes—"something will always turn up." This is the **J-P** scale. The highly organized folk who want to put a grid over life score strongly **J**, while those who are transactionally in dialogue with events score highly **P**.

Putting It All Together

Having completed the Myers-Briggs self-assessment questionnaires, your score will position you on each of the above four axes, combining to give a single, four letter code. The author, for instance, scores as **INTP**.

- ■ I = Introverted (rather than extraverted)
- ■ N = iNtuitive (conceptual rather than concrete)
- ■ T = Thinking (logical rather than values-driven)
- ■ P = Perceiving (understand /persuade rather than dominate).

This is a typical profile for a researcher, architect, scientist, or strategist.

If you enter **INTP** (or any other Myers-Briggs type) into your favorite search engine, a number of profiles will come up. Many assume you know more about Jungian thinking than described here. In Figure 7.1, the 16 Myers-Briggs types are listed based on the scheme suggested in the book by Isabel Myers Briggs (Briggs Myers, and Myers 1995). Beneath the four letter code, I have included the descriptive word David Keirsey (1998) suggests for roles of this type. So in the top left hand corner, we see the dominant type in the police and military, the **ISTJ**: Introverted (not too showy) + Sensing (concrete rather than abstract) + Thinking (tough-minded, duty-oriented) + Judging (authoritarian rather than permissive). David Keirsey calls this kind of person an INSPECTOR.

I want to emphasize that the theory behind the MBTI is a lot deeper than I have described. For example, the four axes are not really in the same category. The S-N and T-F axes are fundamental, and the E-I and J-P dimensions modify

ISTJ	ISFJ	INFJ	INTJ
Inspector	Protector	Counsellor	Mastermind
ISTP	ISFP	INFP	INTP
Crafter	Composer	Healer	Architect
ESTP	ESFP	ENFP	ENTP
Promoter	Performer	Champion	Inventor
ESTJ	ESFJ	ENFJ	ENTJ
Supervisor	Provider	Teacher	Field-marshal

Figure 7.1   The Sixteen Types.

Table 7.1 The Four Keirsey Temperaments

Guardian (SJ)	Artisan (SP)
Idealist (NF)	Rational (NT)

their dominance and orientation. And there is sometimes a facile identification of sometimes-desirable traits such as *conscientiousness* with a single MBTI dimension such as "more J than P." Thankfully people are not so simple, and the overall personality type is not a simple additive combination of the separate letters. To probe further, see Kroeger, and Thuesen 1989; Briggs Myers, and Myers, 1995; Thomson 1998 in that order. If you are interested in knowing your own type, Keirsey 1998 has a sample questionnaire. For more sophisticated insights, a standardized psychometric test set given by a qualified practitioner is necessary.

The Jungian tradition underlying the Myers-Briggs approach appear to some people to presuppose a kind of functional architecture for personality. Not everyone wants to go that way, and David Keirsey for instance took the four Myers-Briggs dimensions and the 16 types but interpreted them differently. Best known for his book *Please Understand Me II* (1998), Keirsey reviewed the history of personality assessment, arguing that there are fundamentally four human temperaments, a view going back to antiquity. Keirsey renamed the four temperaments and identified them with Myers-Briggs dimensions as in Table 7.1.

- **Guardians** are SJs, concrete and institutionally oriented.
- **Artisans** are SPs, concrete and individual-action oriented.
- **Idealists** are NFs, conceptual and value-driven.
- **Rationals** are NTs, conceptual and logic-driven.

The popularity of Keirsey's views arises from the obvious fact that most people are fairly easily seen to fit into one of these four categories, and that it is easy to make finer-grained distinctions if necessary using the full 16 types, which are embedded within Keirsey's approach (Table 7.2).

Keirsey provides distinctive profiles for the four temperaments (1998), as well as the finer-grained types within them, which seem to many people to be both insightful and to have strong predictive powers in both interpersonal and career contexts.

Table 7.2 The Keirsey Temperaments and Myers-Briggs Types

Guardian	ISTJ	ESTJ	ISFJ	ESFJ
Artisan	ISTP	ESTP	ISFP	ESFP
Idealist	INFP	ENFP	INFJ	ENFJ
Rational	INTJ	ENTJ	INTP	ENTP

Given that Myers-Briggs' **SJ**s are Keirsey's **Guardians**, the priest would have been more accurate in saying, "it sounds a bit like David Keirsey, doesn't it" (confusion in the congregation). In fact, the Catholic Church is remarkably keen on personality profiling, especially the Jesuits (the *Society of Jesus* or in Latin, *Societas Jesu*—SJ), and the Jesuits are widely seen as a **Rational** (NT) Order, not **Guardian**.

Personality Assessment in Practice

There are plenty of personality assessment tests, but the classification schemes can be used without them. Most of us think we can judge people pretty accurately once we know them a little. Personality type is a fundamentally *empirical* framework for organizing those judgments, and it is perfectly possible to make a judgement about someone's type by observation. Once we have the right category, we can think more clearly about that individual's personal style (which translates to strengths and weaknesses in a corporate context). That, after all, is the purpose of personality assessment and psychometric testing in the first place. Needless to say, we should base definitive management decisions on formal testing rather than impressions—the error-rate is far lower.

Both Alicia and Mark were Rationals, although Alicia was an introverted, perceiving Rational—**INTP**, while Mark was an extraverted, judging Rational—**ENTJ**. Piers was a Guardian, an extraverted and friendly **ESFJ**.

By reading type descriptions in the books cited in this chapter, or by entering a four letter Myers-Briggs designator, such as **ENTJ**, into a search engine, it is possible to refine your understanding of how different kinds of people are likely to operate, and what the differences are. Sales people have found this particularly valuable. It is unfortunately very easy to fall into the trap of interacting with other people within your own preferred typological style.

- If you are a **Rational**, you will tend to embark upon a long and logically-detailed explanation almost as soon as you open your mouth—lecture or problem-solving mode.
- If you are a **Guardian**, you will tend to address the pleasantries to break the ice before getting down to business, and then focus on the immediate details.
- If you are an **Artisan**, you will be more focused on the immediacies—less social overhead and no abstractions.
- If you are an **Idealist**, ethics and values will always be lying beneath what is said.

I recall meeting recently with a client, who was the technical director of a large company. We were facilitating his involvement with another of our clients. The purpose of my meeting was to understand better his company's approach to Voice over IP. Having found the premises and his office, I assumed he was a technologist (typically a **Rational**) and, after a few preliminaries, launched right in.

Me: Are you just SIP-based or do you also do H.323?

Client: No, that's not it.

Me: I mean, do you have some kind of architecture diagram I could look at?

Client: Look, I don't think so.

At this point, I noticed that the client was looking very defensive, with his arms drawn across his body. He seemed uncomfortable. I blundered on for a few more technical questions until the client lost patience.

Client: Could you explain why you are here?

Well, I thought I had covered that in my first couple of sentences, but now I get it. This guy is *not* a **Rational**, he's a **Guardian**. He doesn't do "straight into technical," we have to get acquainted first and set the scene before doing business. So, I reverse up, explain the background, explain my role, ask him about his history and establish some context and common ground. Finally, after ten minutes or so, the client loosens up, and begins to take me—in his own way—around the topics he has decided I should know about.

Rather than just hitting our conversational partner with our preferred style, we are more successful if we adopt the preferred style of the person we are talking to, or at least adjust to it. And that depends upon identifying their type. Most people with some degree of empathy have the ability to instinctively adjust their style to the people they're dealing with, but not everyone in business is naturally either diplomatic or empathic, and type theory is there to help.

Applying temperament or type theory in an informal setting is useful and will repay the time taken to learn it. The other main use is in more formal personnel selection. It is usually possible to identify the type characteristics of a role, either by the nature of the role, or by examining the types of the people who are conspicuously successful in it. It is not always the case, of course, that one role = one type. For example, a classic police/military pairing is the Guardian with the Artisan. The Guardian is orthodox, and does it by the book; the Artisan is a maverick who does what it takes, paying lip service to the rules where necessary. Choose your favorite film. Each style work best in some situations.

Isabel Myers profiled a sample of urban police with the following results (Briggs Myers and Myers 1995, 50).

- Guardian—56 percent
- Artisan—24 percent
- Idealist—8 percent
- Rational—12 percent

Against general population figures (roughly 38 percent Guardian/Artisan, 12 percent Rational/Idealist; Wideman 1998), this proportion is significantly skewed towards Guardian. But even so, there were apparently ways of being a police officer—hopefully successful—that could exploit Idealist and Rational temperaments.

As we saw in the examples above, there is not one, interchangeable job called "being a line manager." Some departments need strong, task-oriented leadership, some need leaders who care. Sometimes change needs to be accomplished without taking prisoners; sometimes a department, damaged by endless change and uncertainty, needs to be recreated as an effective team. Type awareness, both in defining role characteristics and in assessing candidates for roles, is a tool that is available, highly useful, and, it seems perverse to ignore it.

If line management is the glue that keeps an organization together and provides the recurrent revenues, then it is projects and programs that are the mechanisms of change. How does anything get changed in an organization? You have to establish a project. And, of course, a standard cliché in business is how often projects fail. There are many standard lists of why projects fail: lack of defined methodology, lack of clarity on project objectives, continual change in requirements, lack of empowerment for project leaders, and so on.

For strategic change, we prefer to talk about programs. A program is large, will significantly change the organization, is more likely to be transformative than sustaining, and will spawn a number of projects under its umbrella. Programs come with some well-defined roles at the top, and the correct mapping of people to roles is another mission-critical differentiator between successful, and unsuccessful programs.

Roles, People, and Successful Programs

We need to say more about the difference between programs that are sustaining, and programs that are transformational. A sustaining program is one that accomplishes something that has already been accomplished many times before. The organization understands most, or all aspects of the process and the desired result. These kinds of programs need only *effective* management to secure success. The diverse customers of the program will shout if anything goes wrong, and the program manager will make the necessary adjustments and conduct fine tuning. Sustaining programs can be carried out by "average people executing good processes."

Transformational programs by contrast, are designed to bring about a new outcome that the organization may not, at the outset, fully understand. A transformational program needs *active direction* as well as *effective execution*. To bring about anything really new, you really *do* need good people. The *effective execution* calls for a high-caliber program manager and the *active direction* calls for a high-caliber technical architect and business-process architect. We will now discuss these roles in more details.

Technical Architect

The architecture of a solution is the definition of its components, the relationships between them and the overall integrity of the design and implementation that makes it fit for purpose. In car terms, it's what makes a Rolls-Royce or a Porsche an excellent car, rather than something that ended up with a racing car engine, a tractor chassis, and the body of a bus.

Since both suppliers and customers of a program normally represent special interests and partial points of view, and as specific requirements are always changing, the natural fate of any program, even if well-conceived at the start, is divergence. If this is left uncorrected, the end-point of a program comprises a lot of good work in detail that fails to cohere overall and is unusable, poorly integrated or unaffordable. In a word, a failure.

The technical architect is the design authority for the program. He has the last word on—and final approval of—what is done and what is not done. He is the person who is accountable for the overall design and implementation, and the person who should be able to explain the whole content of the program to any audience. He fights entropy.

In a big program, the technical architect can have a team of specialist architects/ designers under him. In a small program, perhaps only one or two, or the architect may have technical skills allowing him to manage all technical aspects of the program himself.

Normally, the technical architect reports (for program purposes) to the program manager. This is because the organization normally wants the program carried out *effectively* in terms of functionality, time and budget, for which the program manager is accountable. The technical architect's function is an interior one—to keep the content of the program coherent and on-track. Rarely, the organization doesn't really know what it wants, and the technical architect is plugged into the ongoing business discussion and serves to lead and guide the path of the project technically as the business mission evolves. In this case, the program manager works for the technical architect as his enforcer and executive officer. We see this in the military too, sometimes.

The technical architect is not just a stereotypical back-office technical guy. He is responsible for the ongoing "fitness-for-purpose" of the content of the program and will find himself dealing with every type of customer of the program, all of whom will have vital needs that serve to pull the program off-track:

- External customers with new service requirements.
- Internal operations people with process concerns.
- Suppliers pushing their own solutions.
- Marketing people with new product ideas.
- IT people who need to know what they are managing/interfacing to.
- Owners of legacy equipment who need to know how to interface to the new systems.
- Management who need to know what they're getting, and why.
- Endless change requests from each of the above that have to be assessed and approved or turned down.

The technical architect has a number of deliverables—architecture documents, design documents, implementation documents, interfaces specifications, test specifications, roadmaps. The *products* that the system enables are someone else's responsibility, but the technical architect has to specify exactly what the new system will do, and how it can be made to do it.

Business Process Architect

What images go though one's mind when one thinks of business process engineering (BPE)? Most likely a montage of process flowcharts, ineffectual workshops, high-priced consultants, and a business fad that has come and gone. You would be safe to conclude that it is difficult (Harmon 2003).

Increasingly IT departments have been picking up BPE as a core competence. After all, those processes embedded within automation are already in their jurisdiction. But unlike processes encapsulated in legacy computer programs, processes embodied in people's current roles cannot simply be turned off. People matter, whether their fate is to be laid-off or to be retrained. And in the inevitable absence of all useful process documentation, they are often the only source of knowledge as to what actually happens now, a vital input to the process of systems analysis—the first phase of effective BPE.

The target solution that the transformational program has been set up to achieve will be a composite of new IT systems, new technology platforms and new human roles and processes, working at the edge into an existing legacy environment. The business process architect is responsible for developing effective new process models and winning their acceptance by all parties through to final implementation. The skills include:

- Analytical abilities—rapidly getting up to speed on the existing processes through interviews, observation, and document analysis. Understanding the potential of the new systems and technologies. Being able to model the old and new processes, and charting a route between them.

- Technical depth—understanding exactly what the IT systems and technology platforms can do, how they can be used, and contributing to their usability .

- People/political competence—acquiring information, managing workshops, eliciting support, facilitating problem-solving with the staff involved with both the old and new processes, demonstrating empathy while maintaining the integrity of the overall mission.

- Systems/integrative talent—integrating a wealth of different kinds of hard and soft data into a coherent, executable plan and holding the vision.

It may be thought that this diversity of skills is too much for any one individual, and they do tend to pull in different directions, but skilled practitioners do exist. They are part systems analyst and part agent of change. Their deliverables include process models, user-interface specifications, user acceptance test models, and they have a sizeable input into the overall design of the automation systems. They will also contribute to the design of the future organization, work closely with HR people, and take a lead in pilot implementations and making the organizational change happen.

Program Manager

The program manager is responsible for bringing the program to a successful conclusion. The task is mostly about effective leadership. Knowing how to bring about change, who to speak to, how to exert pressure and break through roadblocks. When to use charm and when to use threats. The ideal program manager is tough, resourceful, and can improvise. They have an instinctive feel for how far they can push people out of their comfort zones to achieve mission objectives.

Project managers, by contrast, tend to have more of an administrative bent and tend to be detail people. This is the domain of the Gant charts, the ticking off of boxes, the detailed measurements of progress against plan, and the endless reminders, cajoling and checks. Good project managers are methodical, detailed, and conscientious. A program manager needs project managers to do his leg work, and to act as his eyes and ears as he steers and coaxes the program to success.

The program manager marshals his assistants. The technical and business process architects have already been mentioned, and there will also be finance people managing the budget, commercial people managing vendor negotiations, and any other specialists needed for particular tasks, for example HR if there is a broad people dimension, facilities if buildings and leases have to be addressed.

The program manager has borrowed power. Although as an individual he reports to someone, for the purposes of the program, he is responsible to the executive sponsoring the program, and borrows that executive's power to get things done. For transformational programs to succeed, it is imperative that the program reports at a senior level. In many cases, anything under the CEO or COO is not good enough.

Type Requirements for Program Roles

Using Keirsey's temperament categories, we can draw up a transformational program role correspondence as shown in Table 7.3.

The technical architect and the business process architect roles are not necessarily type-identical. The business process architect would benefit from being more extraverted and less judgmental, as judgmental often comes across as abrasive and arrogant. In Myers-Briggs terms the optimal role definition is probably an affable ENTP who can *listen*. The technical architect role, by contrast, probably works with any flavor of Rational. Why not an Idealist for business process architect? Idealists are people-oriented and driven by moral values; business transformational programs by contrast are focused on task and mission accomplishment, where sometimes people's feelings, and even their livelihoods, get damaged. The program needs to priorities not *interpersonal harmony*, but *effective human resource management*: they are not quite the same thing.

The program manager role has a focus on the effective use of institutions and people, areas where task-oriented Guardians and Artisans excel. If the project requires forcing through, then the Artisan style of task-focused power can be more successful. If it requires extensive use of formal institutional power, then a Guardian can be very effective. Subordinate project management functions favor Guardian types to the extent that they require routine conscientiousness over unconventional problem-resolution skills.

Table 7.3   Program Roles and Individual Temperaments

Role	Temperament
Technical Architect	Rational
Business Process Architect	Rational
Program Manager	Artisan/Guardian
Project Manager	Guardian/Artisan

It is important to recall that this discussion is indicative, not mechanical. Almost all types can be successful under certain kinds of circumstances and with appropriate support. But selecting people for roles ought not to be an arbitrary process, and the conceptual framework discussed here is best-in-class for assessing both role needs and personnel suitability. The next chapter gives an example of the impact of personality type on program success.

High-Performance People?

Every large organization I have worked in has had a program for High-Performance People (HPPs). There is some kind of evaluation process, by which the HPPs (sometimes called "hypers") are identified and logged. Their careers are then meant to be managed separately, putting them on a fast-track to senior positions later. I think it is fair to say that every such program I have ever seen has failed, and after an initial burst of enthusiasm, the scheme falls into abeyance.

- No one knows what to do with the HPPs.
- There is a reorganization and the HPP program gets lost.
- People advance their protégés, and ignore the HPP database.
- HPPs are valuable to their current managers and are not let go.
- HPP database information is inaccessible, too limited, inadequate, or incomprehensible.
- The whole scheme is loathed by the majority of non-HPPs, including many managers.

People are understandably cynical about such programs, but the underlying need to identify, develop and use talent is undeniable. Since people are corporate resources, there is really little excuse for line managers not understanding the concepts of personality type (one reason for this chapter). Personality/IQ assessment of staff needs to be generalized. After all, if job candidates can routinely take a battery of IQ and psychometric tests, there is no reason why staff, who have already been employed, should not be able to do likewise to the advantage of themselves for career development, and of their managers for resource development and allocation.

On this basis, skills and talent management would be part of every line manager's day job, rather than being hived off to a separate program. All that needs to be added is a regular upward submission through the management line of top performers who are candidates for development. Without the structural separation between HPPs and everyone else, morale and team cohesion also improves.

Summary

In this chapter we have taken a look at individual differences in personality characteristics and described various frameworks and terms for capturing distinctions *explicitly* that we already *intuitively* know. I have argued for more intensive use of psychometric testing to create a more accurate and objective basis for optimal staff allocation and career development.

I have devoted a large amount of space to personality testing and not much to intelligence testing. Any psychometric testing suite will include both kinds of assessment. It has been known for decades that the ability to function at higher executive levels directly correlates with measured IQ, but that effectiveness in-place is then determined by the nature of the role and the personality characteristics of the individual concerned. Raw aggression can only get you so far before the executive coach is called in. I am an advocate of intelligence testing in business personnel assessment because IQ is a robust predictor of potential performance, other attributes being equal.

We looked in some detail at the challenges of transformation programs, and the roles and skills needed to make them successful. In the next chapter, we take a look at a case where these concepts will prove valuable in understanding what actually happened.

In North America, I would say that the concepts of this chapter are well-understood in most leading companies, and that those companies achieve competitive advantage through their use. Outside of North America, understanding and take-up is considerably more patchy, and I would urge managers to be pro-active in bringing these ideas into their daily work.

Appendix 7.1. The Scientific Take on Personality Classification

This appendix is for those who wish to know more about how the scientific community thinks about personality differences, and why the Myers-Briggs approach is not at all the end of the story. It can be skipped without any breaks in continuity.

While most corporate HR staff and many therapists and counselors routinely use the Myers-Briggs and Keirsey assessments, the academic community analyses personality in terms of the Five-Factor Model (FFM) (McCrae and Costa 2003) comprising the traits of: Openness, Conscientiousness, Extraversion, Agreeableness, and Neuroticism; sometimes recalled through the mnemonic of **OCEAN**. This is well described by Pierce and Jane Howard (2004), who run one of the few consultancies using the FFM in a business context with clients.

The FFM dimensions arose from the following approach. Personality is important in human relationships, so we are likely to have a rich vocabulary of significant personality terms—this is called the lexical hypothesis. Large numbers of personality words were therefore culled from dictionaries, and people were asked to describe other people using these terms. Then a technique known as factor analysis (Kim and Mueller 1979) was used to group terms that seemed to correlate with each other, and therefore might mean the same thing. After much number-crunching and discussion, it seemed that personality terms seemed to cluster predominantly around five largely-orthogonal major axes: these constitute the Five-Factor Model. Naturally, there is much debate about personality concepts that don't seem to fit the model—honesty?—but a consensus seems to have developed.

Note that unlike Myers-Briggs and Keirsey, there are *no types* in the FFM. The five axes constitute a kind of five-dimensional personality space, and an individual taking an FFM personality assessment ends up with an aggregated score along each dimension, a personality vector in this space. In practice, each trait is further subdivided into six subtraits that are also chosen to minimize cross-correlation; the resulting raw personality vector has 30 components. The significance of this collection of numbers can be hard to understand—the typological analysis, once the letter codes are understood, seems easier to work with, which is probably why it is more popular in a corporate or clinical context. People who have used the FFM with clients have scored the assessment in terms of High, Median, or Low along each of the five main dimensions. This is a little more manageable, but you end up with $3^5 = 243$ possible outcomes, rather than 16 types.

Academics don't like the Jungian apparatus of dominant, auxiliary, tertiary, and inferior functions and their attitudes, and they don't like the apparently bimodal character of the type attributes—where you are forced to choose between alternatives: **E** or **I**, **S** or **N**, **T** or **F**, **J** or **P**. In the Five-Factor Model, populations are assumed to distribute normally along the five axes. Since both approaches are agnostic about the underlying neurological architecture of personality, the differences in neurological structure and function that underlie observable personality differences, and are not situated within an evolutionary paradigm, this is clearly an area where future research has the potential to transform current thinking. I am inclined to appreciate the insights afforded by the Jungian approach, without thinking of it at all as the last word—a kind of psychological analogue to Aristotelian or Newtonian mechanics.

In fact, it is possible to compare the Myers-Briggs (and by extension, the Keirseyan) approaches and the Five-Factor Model. The Wikipedia article on MBTI (http://en.wikipedia.org/wiki/MBTI) quotes a study of 119 undergraduates who were given both the Myers-Briggs assessments and an FFM personality inventory and the results compared to look for correlations (Figure 7.2).

	Extraversion	Openness	Agreeableness	Conscientiousness	Emotional Stability
E-I	**65%**	6%	-37%	-15%	31%
S-N	12%	**-56%**	34%	37%	6%
T-F	19%	-25%	-21%	9%	7%
J-P	18%	-15%	10%	**55%**	8%

Figure 7.2   Correlation between FFM traits and Myers-Briggs dimensions.

Any correlation over 50 percent can be considered significant, so what this is saying is that:

■ Extraversion (FFM) seems to be extraversion (**E**) on the **E-I** dimension.

■ Openness (to experience) is similar to iNtuition on the **S-N** dimension.

■ Agreeableness seems to be a composite in the Myers-Briggs view: you are perceived to be more agreeable if you are somewhat introverted (**I**), somewhat concrete rather than abstract (**S**) and to a lesser extent, somewhat Feeling rather than Thinking (**F**). Perhaps there is more than one way to be agreeable! However, other FFM theorists have simply identified the Myers-Briggs **F** function with Agreeableness (McCrae and Costa 2003, 56).

■ Conscientiousness correlates with **J** on the **J-P** dimension.

■ Neuroticism (alert, anxious, worried) and its inverse, Emotional Stability (calm, relaxed, stable) does not seem to be captured in the Myers-Briggs approach.

The Myers-Briggs testing found in corporate life is focused more on cool, top-level psychological functions than on the overall psychological integration of these with deeper emotional drives. Interestingly, it was the latter that was the focus of Jung's mainstream work in analytic psychology. In a business context, low-neuroticism, sometimes described as high emotional stability, is usually a required personality attribute (except, perhaps, for some of the most senior managers).

There is a mathematical way of thinking about this. Consider the totality of a person's personality to be represented as a vector **P** in some high-dimensional space as yet not completely understood. Then both Myers-Briggs and the Five-Factor Model are *representations* (sets of basis vectors, but not complete sets) onto which **P** can be projected. Some of the axes line up (*Extraversion* with *E-I, Openness* with **N-S**) while others are linear combinations (*Agreeableness* might be a linear combination of **E-I, S-N, T-F**). The extra Five-Factor Model dimension of *Emotional Stability* is orthogonal to the Myers-Briggs set of bases. On this view, fights between the FFM and Myers-Briggs are arguments about the *utility* of different representations, not about the fundamentals, since they are related by linear transformations.

Recent work in Evolutionary Psychology suggests that the Five-Factor Model axes should be rotated in personality space to align better with evolutionary-significant traits. In this view, the two orthogonal axes of Extraversion—Agreeableness should be rotated to Dominance—Nurturing. Expected gender differences then appear (MacDonald 2005, 210). Anthony Stevens and John Price (2000) suggest that human beings rank along two orthogonal dimensions: dominance (rule by fear) and social attractiveness (rule by attraction/admiration). They note that most people prefer to operate in institutions governed by the second model, which is normative in most business and social environments (cf. the perennial search for the team player). The criteria for high-ranking on the social admiration axis can be varied: athletic skill, intelligence, personal effectiveness, a genial personal style, moral courage. Of course, rule by fear on the first axis is not by any means unknown in business.

Incidentally, as a reader of this, you are most likely to be a **Rational**. Failing that, I expect some **Guardians** as readers, looking to properly prepare themselves for the NGN: **Guardians** are often interested in technical matters and tend to be conscientious in reading everything relevant. You are unlikely to be an **Idealist**, not only because there are so few in telecoms, media, and technology, but also because their people orientation tends to steer them away from books like this. And finally, **Artisans** do not read this kind of book, ever (apologies if you are the exception—someone probably made you do it).

References

Briggs Myers, I., and Myers, P. 1995. *Gifts differing: Understanding personality type*, Mountain View, CA: Davies-Black Publishing.

Harmon, P. 2003. *Business process change*, San Francisco: Morgan Kaufmann.

Howard, P. J., and Howard, J. M. 2004. *An introduction to the five-factor model of personality*. Center for Applied Cognitive Studies, http://www.centacs.com/quickstart.htm.

Keirsey, D. 1998. *Please understand me II*, Del Mar, CA: Prometheus Nemesis Book Co.

Kim, J. O. and Mueller, C. W. 1979. *Introduction to factor analysis.* Newbury Park, CA: Sage Publications.

Kroeger, O. and Thuesen, J. M., 1989. *Type talk.* New York: Bantam.

McCrae, R. R., and Costa, P. T. 2003. *Personality in adulthood—A five-factor theory perspective* (2nd ed.). New York: Guilford.

MacDonald, K. 2005. Personality, evolution and development. In *Evolutionary perspectives on human development* (2nd ed.), eds. R. L. Burgess, and K. MacDonald, chapter 8. Newbury Park, CA: Sage Publications.

Stevens, A., and Price, J. 2000. *Evolutionary psychiatry: A new beginning* (2nd ed.). Hove, East Sussex: Brunner-Routledge.

Thomson, L.. 1998. *Personality type: An owner's manual,* Boston: Shambhala Publications.

Wideman, R. M. 1998. Project teamwork, personality profiles and the population at large, Proceedings. of the 29th Annual Project Management Institute Seminar/Symposium "Tides of Change," Long Beach, California, (updated presentation, April, 2002), http://www.maxwideman.com/papers/profiles/observations.htm.

Chapter 8

Case Study: A Transformation Program

Introduction

In the previous chapter, we discussed transformation programs and the selection of individuals for the key roles within them. This task is more difficult when a major program is outsourced to a consultancy. The consultancy may lack fine-grained knowledge of the host organization's culture, and a successful model in one environment may be hard to replicate in another.

These points were particularly relevant when a consultancy I was associated with was asked to lead a large IPTV-VoD (Video-on-Demand) program for a European company, JDIcom. The relevant technology was discussed in chapter 3. The consultancy had had little to do with JDIcom previously, all we knew about it was that it was in the networking and content distribution business, and that it was eager to exploit new opportunities for voice and video over IP to residential customers. The company was multinational in its staffing, but the working language was English.

JDIcom was used to running programs, but typically only one major program at a time. As it had grown, and as the rate of industry innovation had increased, it was having to run more priority programs in parallel. Senior management had decided that a more formal project management discipline was required, and that an outside project management consultancy might be the way to import it. The team was comprised of me and two colleagues, George and Peter.

George had been assigned the program director role. He had previously been a senior operations executive with a major carrier. He was a big, burly man with a generally genial, paternal manner, although a certain directness sometimes featured in conversation with him, where another might have chosen a softer way. Put George in a program management situation and it was all business: the bonhomie was still there for situations he approved of, but if you irritated him by failing to meet your obligations George could crank up a level of intimidation you would not believe. Later, in meetings, I would see George turn his powers of persuasion on recalcitrant project managers, and it was as if an implacable black wall occupied the room, horizon-to-horizon and extending far up into space. This wall of coercion would slowly advance to enclose the hapless individual—resistance truly was futile.

George was familiar with the Myers-Briggs scheme, as discussed in the previous chapter, and later in the project he was happy to confide to me that he was an INTJ. I have to say that I was quite surprised to hear this, as he seemed to me to be a Guardian type: I would have guessed ISTJ.

However, when you take a Myers-Briggs assessment, the facilitator typically tells you the type that the tests have indicated and asks you whether you agree. If you disagree, you are invited to say what type you think you are, and that is the type that is then taken as definitive. From a formal psychometric point of view, this is not a good policy, although it clearly helps user acceptance in client practice. I recalled that when I was working in Nortel, I had believed myself to be INTJ and had indignantly corrected test supervisors who had scored me as INTP. Reading the type descriptions naively, I saw INTPs as detached bystanders, while INTJs got things done. I saw myself as a go-getter and change-agent, very much exemplifying the Nortel results-oriented culture. It was only later I realized that any type can get results, in their own way, if that's what is required, and that personality classification is looking at something far more profound. I had acquired a deeper understanding of what kind of person I was, and the *way* I responded to the challenges of the culture around me.

George was interested in ideas and proud of his academic accomplishments, although he would not have described himself as an intellectual. I would guess that, contra Jungian presumptions of necessary bi-modality, George was probably on the S-N boundary. Just speculation on my part.

My other colleague, Peter, was physically smaller and thinner, and much more the classic intellectual. Where George projected visceral authority, Peter brought understanding and penetration, a powerful motivation for order and detail, plus a relentless determination to get to the bottom of all issues that might affect program success. Peter was not familiar with the Myers-Briggs scheme and had not taken any assessments. To be honest, he was not that interested, but once the typology was outlined, he readily agreed he was probably a Rational INTJ.

In their self-contained manner, tough-minded approach and propensity to organize anything that moved, George and Peter matched each other perfectly. Peter produced ideas, and George supplied common sense. They both majored in conviction.

I was the odd one out here and I wondered why I had been asked to join the team. As a Rational INTP, I was useful for ideas and strategy, and I happened to know quite a bit about IT from previous work and assignments. Being **P** rather than **J** registered a more laid-back, diplomatic, and transactional attitude to getting things done. But in the more robust language of program management, however, I was probably too *passive*, too prone to *go native*, too inclined to *see the other person's point of view* and therefore a *weak sister* when it came to cracking the whip and enforcing project discipline. Guilty as charged, Sir! I have never pretended to be a project or program manager of the old school. JDIcom's problems looked interesting, but how my own role would work out was anyone's guess.

The First Meeting

After a few days to prepare, we were invited to a meeting at the client's site, which brought the stake-holders together and kicked the program off. As the meeting progressed, we came to the following view: there was clearly momentum already for this project, as work was already being done, particularly with JDIcom's hardware, software, and security infrastructure partners. On the other hand, there was no overall mechanism for securing a cohesive solution, for bringing the various projects together and for allowing effective overall management.

This, of course, was not a surprising conclusion as it was precisely why we had been brought in. We made sure that we sent around a sheet of paper so we could get everyone's name, role, and contact details, and subsequently scheduled a series of meetings with all the key players: architects, relevant line managers and the assigned project managers, to begin our detailed engagement with the program.

Over the next few weeks, we began to learn more about JDIcom. Its style of work was osmotic: coordination was achieved by cross-silo involvement in each of the projects. The upside of this approach was the presence of a diversity of skills at each meeting, and the rapid diffusion of information around the organization. The downside, in addition to the size of the meetings, was that everyone of any skill had their time sliced between many different working-groups, and was involved in endless meetings. Apparently, attempts to install more process were seen as harbingers of the dreaded bureaucracy, which would be anathematic to JDIcom's start-up culture.

We have seen the evils of bureaucracy in many carriers (cf. chapter 5) where innovation is all-but-impossible, and never fast, so we enormously sympathized. However, we *do* need process and accountability, and we do need to free up some space for work. As the saying goes, you either meet or you work. After consultation with our senior management contacts, we therefore put forward a proposal for a new, cleaner project organization and reporting structure. We recommended that the program should be divided into a number of projects, each headed up by a project manager accountable to a central program office that we would constitute ourselves.

Program Start-Up

After two or three week's of interviews and document reviews, we reached that most difficult of times, the point where you try to take charge. JDIcom was a most polite culture, and although everyone was incredibly busy, we had had little problem in getting into people's diaries when it came to interviewing them. But soon we sensed our honeymoon period was coming to an end, and it was time to add some value (Figure 8.1).

Our recommendation was accepted. We went ahead and established the program office, comprising George as program director, Peter, a JDIcom finance guy, and me. Wednesday was our first *Program Review Meeting* where we allocated the whole day to meet with the project managers from each of the program's project for an hour each. They had each been asked to prepare a brief report with the following headings.

- Milestones reached
- Remaining milestones to end of program
- Achievements since previous review

Figure 8.1 The consultants.

- Issues/Concerns/ Risks to future milestones
- Decisions Required
- Approvals/Authorizations required

At this stage, we were not expecting too much, as many projects had barely got going, and planning was in its infancy.

First up was Aubrey from the marketing project. Previously, this project had been stalled, as Aubrey had been totally focused on the JDIcom network evolution program. As a result, he had had little time to worry about our IPTV concerns. However, over the last week, a lot of work had been done and the whole area was now shaping up.

Next was Charles, one of the principal technical architects, who specialized on the network side and was closely involved with the network division people. The temporary project manager for the Technical Design Authority (TDA) was away, and the new one had not yet arrived. Charles was the classic architect: smart, laid back, and accommodating. He had the slightly world-weary air of someone who has seen all the stupidity in the world, has given up the futile task of fighting it, and is content to navigate around idiocy and do what he can for progress. God knows what he thought of us. Charles was helpful, but didn't fit our needs for what a project manager was meant to be doing in this slot.

The Set-Top Box project manager was Keith. He was a thirty-something no-nonsense guy who George took to instantly. This had something to do with the fact that Keith had completed his form and showed every sign of understanding what it meant to work to a plan. While the STB area looked in good shape, we had been warned repeatedly that it has the potential to sit on the critical path and cause delays. The box was complex and was the focus point for delivery of the entire service. Often problems only surface late in user trials, when change is difficult and time very short.

The network representative failed to arrive. I had predicted this, as it was always going to be unlikely that the network division's technical director, who is being pulled a thousand ways, would have made the two hour trip from his office to our program office location just to emulate a project manager. The network division was preoccupied with their expansion program, and had not been able to assign a project manager to us. Another problem we will have to deal with.

Harold, leader of the broadcast and systems project, visited us next. Broadcast and systems was tasked to put in place the MPEG encoders that pull channel content into IP and feed it into the network. On the VoD side, they will acquire and integrate the VoD servers and asset management systems, and integrate everything with the Conditional Access System. And then there are the changes to IT systems, both scheduling, EPG, advertising systems, and the more traditional business support systems involving CRM, field-force management and billing. The IT impact at this stage was completely unknown.

Harold confirmed that they had started plan development for the broadcast infrastructure head-end, and that it didn't look good. If they were to do the program the conventional way, they could miss the launch by months, not weeks. They were working on it.

Our final project is Business Readiness. This is Customer Operations plus whatever else is needed to get to launch. We talked to the responsible executive, who, like Aubrey, was totally absorbed with the network expansion program and hadn't given two thoughts to our IPTV-VoD program up to this point. We didn't get mad. We stressed our need for a dedicated project manager and he promises to help. Someone had already been allocated, apparently, and would attend the next meeting.

Finally, we met with Nick, a suave and suited lawyer representing JDIcom's legal division to us. This covers a number of areas: customer contracts, regulation, rights, contacts with suppliers, lawsuits, and so forth. We were told that the area of most relevance to our program was the question of rights: specifically the right to distribute channels across a streaming IP network, and the rights to place material on Catch-Up TV and VoD servers.

Our series of meetings ended around 5 p.m. We felt that at least we now had a calibration of where people were at.

Friday saw our second governance structure, the Program Management Board. This is the somewhat grandiose title for a weekly 90 minute meeting of the Program Office with all the project managers together. The intention was for us to report back on progress, and for the meeting to discuss and resolve issues that cut across multiple projects and that cannot easily be sorted out bilaterally. We presented slides we had put together after Wednesday's PRM, and George gave a pep talk about the need to expedite planning and the need for each project to make an assessment of its state-of-readiness and its level of risk.

Wednesday's PRM and Friday's PMB was the formal side of governance. In an organization used to running programs this way, we would simply start things up, get everyone used to this style of reporting and oversight, and after a few weeks we would expect a "program under management." However, JDIcom is not that company. The culture here, as mentioned previously, is much more organic. People are used to multi-tasking and networking to achieve their objectives. From this perspective, formal procedures can look a lot like mindless bureaucracy, adding a layer of meetings and reporting that is just deadweight.

This view is not stupid. Many media companies, as well as start-ups, work effectively in this informal manner. However, given the number of high-priority projects now running concurrently in JDIcom, a degree of formality is vital if the aggressive dates are to be hit. Our challenge is to add a framework of discipline, structure and formality in a way that helps the project, not hinders it, and to sell the case to the people whose help we will need to make it work. It is not easy.

...e process. The program cannot just grow organically, like a plant at
...tip of the shoot, bending around obstacles as it encounters them.
...milestones for the duration of the program, with dates and dependen-
...s a benchmark solution architecture for the program. Then it can be

...old the whole thing together arguably needs someone like George:
...thoritarian, charismatic even. Someone who can get your attention,
...u will feel uneasy saying "no" to. Almost all of the JDIcom managers
...h is why they adopted their role placements as prescribed in theory
... their skills to manage our expectations. And hence my discussion

...alk with Arthur have any success? He did try to shape up, but you
...ge character. The team came to see him as bureaucratic, someone
...py to organize his own area of responsibility, but who was essentially
...regards other areas of the program. We wanted someone who would
...to pro-actively drive any part of the program if it was necessary to
...wn objectives, not simply manage within the confines of their own
...aps we were unfair. Perhaps we projected onto Arthur some of our
...ncerns about our own performance.

...rks People

...ss problems getting any attention from the network division. They
...of resources, and intense concentration on their new network ex-
...ram. The effect on us was: no plan, no key milestones, no dedicated
...ger and, therefore, a nonexistent project.
...sed a dedicated project manager, a specific individual we knew,
...ejected on the basis that his CV was too operational, and did not
... the network design area. We thought the rejection came a little

...ng week found me at the network division's main site. After the
...he meeting started at 5.30 p.m.—those were the kinds of hours the
...eeping. I was meant to be meeting the network technical director,
...e man named August, with an introduction made by Victor, the
...had been our main contact thus far. However, Victor showed no
...go away.
...g room in which we sat seemed far too big for the three of us.
...ced at one end of a very long table, close by the door. I began by
...own background and indicated the problems we were having in
...network project into the overall IPTV-VoD program. I indicated
...elp overcome the difficulties and to find out exactly what extra

The Executive Board

The following week brought the first meeting of the IPTV-VoD Executive Board which brought together project leaders and relevant senior executives from JDIcom under the chairmanship of the COO. George, as Program Director, attended from the Program Office, and we prepared the meeting.

The Executive Board is the third leg of the program governance structure, along with our innovations of the PRM and the PMB. Executive Boards are a venerable JDIcom institution and their function is well-understood: to allow senior executives to get a snapshot of progress, and to allow issues to be definitively resolved.

The issues here included the number of TV and radio channels to be provided on day one of service launch, the problems of hitting the very aggressive launch dates, and the challenge of scoping how much work needed to be done in the IT space.

Talking to a number of people afterwards about the meeting (which I did not attend) I was struck by how the same meeting can be reported in such a diversity of incompatible ways. Did the IT people get beaten up badly, or was it merely an action item competently addressed and then moved on from? Was it really more of an issues-oriented workshop rather than a senior management progress review meeting? Did senior staff really micromanage down to junior staff levels of detail, and if so, was this really a bad thing? It was impossible to say.

The "Difficult" Project Manager

Arthur first came on the scene as a new project manager. His boss had mentioned his recruitment, and had spoken very highly of him, so we were intrigued to meet with him. Arthur came into our room and we did the introductions. Both George and Peter then had to leave, but as I had nothing planned, Arthur and myself were able to talk for a further hour. It was nothing special—I explained the project, and he talked a little about his previous career. My take-home view? Arthur was bright, opinionated, slightly prickly, but definitely someone we could do business with (probably a Rational ENTJ).

Over the next week, my colleagues begged to differ. I heard that he was difficult and uncooperative, that he was a baleful and negative influence on the program. I could not join in with these sentiments, and was genuinely puzzled: why did my colleagues take against Arthur so? Arthur himself remained oblivious of the impression he was making, just seeming puzzled that a certain tension existed between himself and the Program Office.

The scales fell from my eyes on the next occasion of the weekly Friday PMB meeting. All the project managers were there, along with some of their managers—

quite senior people. George was chairing, and I noticed that even when he was giving them a hard time, the JDIcom managers were quietly respectful of George, trying to help him and themselves in pushing the program along.

Arthur's approach was somewhat different. George, as program director, would suggest something. Arthur would disagree, and, in a manner that might appear to some as condescending, he would point out the drawbacks and suggest some improved notion. (This, by the way, is typical extraverted Rational behavior). If he expected George to instantly see the merits of this and adopt his idea forthwith, he was to be disappointed. Instead, George bristled.

Later that day mulling over these events, I had an abrupt realization. The truth was that Arthur was *insubordinate*: not at all in the insolent sense of the "lad," deliberately flouting authority, but in the manner of the intellectual, who does not even see that the authority dimension exists. By failing to understand and accept the authority relationship between himself and George, he was actively undermining George's authority. No wonder George was taking it badly, Guardians care deeply about that sort of thing. I resolved to confront Arthur for coaching the next time I saw him.

Monday morning arrived, and my good intentions felt a lot less compelling as I approached the office shortly after 8 a.m. It is difficult to raise these issues with people; much more dangerous than merely having technical discussions. As I walked by his desk, Arthur's coat was there, but he wasn't. I left a note. A few minutes later, he turned up and with some trepidation I ushered him into an empty meeting room.

We agreed right away that relationships between my colleagues and himself were strained and needed to be improved. Arthur knew there was a problem, but couldn't understand why.

"Arthur, who do you work for?"

Arthur looked nonplussed at this question. He knew that I knew the answer and named his boss.

"Yes," I said patiently, "who else do you work for?"

Arthur looked puzzled and I had to prompt him.

"George, right? He's the Program Director, and you are a project manager on the program. So in program terms, you report to him and he's your boss. It's a matrix thing."

"Yes, yes of course. I know that."

"Well, you know it and you don't know it. That's the problem."

Arthur stiffened, feeling the need to defend himself.

"What is George's problem? I am trying to help, I listen to what he says and I try to respond constructively and to give him the benefits of my experience and ideas."

"Yes, Arthur, that's very apparent. However, the problem is not the content of your ideas, which are good, not even your readiness to contribute, also good. The problem is with your attitude. To put it bluntly, you are insubordinate."

At this point, Arthur showed the classic sy[...] the words, but did not process their meanin[...] propriate.

"Well, what does George want? If he does[...] stop giving them."

We continued a dialogue of the deaf for s[...] the point that this was not a question of an [...] propriate authority-behavior— knowing yo[...] nowhere, I had to change tack.

"Arthur, I've tried the sophisticated stuff[...] me try the tabloid way. Suppose this progr[...] the Godfather, the *capo di tutti capi*. Then y[...] showing no respect. Now do you get it?"

Arthur showed no evidence that he belie[...] his own situation, but he did promise to g[...] meeting pessimistic about the results: changi[...] a conscious and sustained personal belief in [...] that Arthur had really bought the argumen[...]

Theory X vs. Theory Y

Let's try to frame this episode. Theory X v[...] Do we apply firm hierarchical control to w[...] and keep them from straying? Or do we [...] conditions so that self-active problem-sol[...] effect?

JDIcom looks to me like a theory Y co[...] hard-working. They know how to collabo[...] to get decisions made when they need t[...] the same way, often being drawn into qui[...] attention is grabbed.

Program management, as George and [...] People are assumed to be wandering from [...] definite dates, precise descriptions of deli[...] ability from individuals who are presume[...] hold them to account. Echoes of the ste[...]

It would be easy to roll your eyes an[...] from the dinosaurs. Surely we are all th[...] knowledge workers? We all know that e[...] reality, and that prematurely imposing a[...] off the creativity and flexibility needed [...]

And yet... to deliver this program to[...]

resources they might need. I even cracked the standard gag ("I'm from the government, I'm here to help you"). I was not prepared for what happened next, as Victor cut across my explanations.

"It's impossible for us to help you when you give us no information about what you want."

"Victor, that's not really the case. It's true that marketing started late, but they *have* made rapid progress recently, and the requirement for IPTV and VOD is now quite anchored down, as ..."

"Well," interrupted Victor, "that's not what I'm hearing from the executives I talk to. It seems there's hardly two of them who have the same opinion."

I struggled to continue. "The Executive Board is authoritative, and it's chaired by the COO. You are a member, and had you been at the first meeting, you would have ..."

But Victor would not be deflected. "I have seen *no* signed-off requirements yet from your program, and when I talked to the CEO about it I'm afraid I got a rather different story. So I would say to you, get your requirements sorted out before you ask us for specific help. We made it quite clear that we could not do your program until we had completed our network upgrade."

I tried again. "The IPTV-VoD dates are quite clear, and if we can't make them, we need to go back to the executive with dates when we *can* deliver, or with a scaled-back proposition that can hit their required dates. As far as I can see, putting QoS and multicast for IPTV over your network would not be too difficult, since the equipment you are ordering will have all the capabilities."

At this point, Victor and August exchanged knowing glances at my naiveté, explaining mock-patiently the difficulties in configuring, tuning, and testing these functions.

The conversation, if it could be called that, continued with attacks on the our program's apparent inability to exactly specify the demographics of intended customers down to the zip-code level, the inability to specify POP locations for VoD servers and their power and heat-dissipation requirements, and the administrative load imposed by our all-too-numerous program meetings.

I once had a girlfriend whose tactic in arguments was to keep changing the subject before I had a chance to properly engage with it. As the goal-posts were abruptly moved time after time, I had felt like a dog being yanked on a lead, or a bull vainly charging a flourished cape ... over and over again.

Talking to Victor, I was back there again, wrong-footed and gamely ineffectual. With the wisdom of years, I could now better recognize what was going on. I was doggedly assuming that the conversation was about how to resolve all these problems. My conversational partner, however, had no desire at all to cooperate: it was merely an exercise in stonewalling.

August had scarcely got a word in edgeways until, at 6.40, p.m., Victor had run out of steam. With his leaving for an evening engagement, I was able to snatch a few minutes with August, who briefly outlined a number of issues with upgrading his network expansion program to accommodate our program. His plan had no slack in it for a TV service overlay activity, there were issues in the access network, which was running on leased circuits not dimensioned for IPTV, and he was resource-constrained in terms of people and skills.

I left the networks building with the view that this was a show-stopper for delivering against the current target dates, and that the level of noncooperation shown meant that we had no obvious means of resolving the situation. In program management, you never give up—there is always a move you can make. But this did not seem to admit of any fast or even effectual response.

Rescheduled

Events continued to move fast, and two days later we were surprised and relieved to hear that the Executive Board had decided to delay our program. The initial planning work from the head-end project had already indicated that this was going to be necessary, and network division's situation had led to an identical conclusion. With the new dates, the program had now become achievable. As I absorbed the news, I wondered whether Victor had known this all along, and had seen me as just another middle manager: sent on an impossible mission in the dark.

On Friday, I paid a second visit to the network division. This time, I met with August and two of his designers, with no Victor in sight. The meeting was extraordinarily productive. It turned out that the network division people knew essentially nothing about our program. Once I had had a chance to brief them, we were able to agree some go-forward plans. Networks were intending to recruit a project manager dedicated to IPTV over the next two to three weeks, and that person would ensure divisional resources would be allocated to our program and the work aligned. With the new date, they were sure they could now meet our needs. A good result.

The final outcome was anticlimatic. There was an internal reorganization in JDIcom and our sponsor changed roles. With the new date, the program didn't seem quite so urgent, and the need for our involvement seemed less compelling. On that basis, our mission with JDIcom was accomplished and we moved on to other work.

Conclusions

In retrospect I think we were too authoritarian. The IPTV-VoD program was truly very volatile, in an environment of other programs that we did not control and of which we had little visibility. By trying to anchor things down early and hard, we created almost impossible pressures on our project managers—pressures we often didn't understand because they were not immediately visible to us.

Being more politically aware would, however, have been difficult. We were not employees and were not part of the extended social network. People were very busy and would not have welcomed us wandering around, randomly engaging them in conversation. JDIcom tried to run one program in an orthodox manner in a sea of other programs—some of higher priority—that were running more informally. This was always going to be a difficult play to get exactly right.

However, JDIcom's fundamental problem has not gone away. Their astonishing ability to innovate is based on hyper-activism on the part of their staff, and a bottom-up culture of best-practice and improvisation. Their good program managers use good methodologies, leveraged from earlier successful programs, the less good ones seem less in control, but there are no standards as such. It seems to me a nontrivial problem to develop a program management template for JDIcom which would exactly suit them. It would have to leverage their creativity, flexibility and intense social networking, while adding the necessary degree of process standardization and formality. And the same dilemma would apply to any other successful, creative and innovative organization.

I don't believe our team ever really understood that point. We imported a model that would have worked with people who were less educated and more used to being led, but that clashed with the culture around us. In the end, the problems probably outweighed any benefits we brought. I am conscious that in saying this I speak as the non-program manager. I suspect my colleagues would have very different views.

BUSINESS AND TECHNOLOGY ISSUES

Chapter 9

Worrying about Skype

Introduction

Skype is a well-known free voice over IP service that has picked up millions of users within the last few years. Unlike many of its competitors, it has achieved almost household name status. Its iconic rank was made complete through its acquisition by eBay for $2.6 billion in September 2005.

In its original, and still dominant mode, the Skype client runs on a personal computer. The Skype executable can be downloaded (http://www.skype.com) and installed on a Windows or Linux machine. The client (Figure 9.1) is protected by user id and password (which can be stored by the client to avoid typing it on every use) and people you contact with Skype can be added to your "buddy list" ("buddy list" is an AOL trademark) where you can double-click their entries to establish a call.

To make or receive a Skype call, you need to purchase a headset with earphones and microphone. A PC's built-in speakers and microphone would also work at some level of sound quality at the expense of privacy and annoyance to others. Skype can also do video-telephony (you need to purchase a Web cam), and provides facilities for instant messaging and file transfer. All communications are end-to-end encrypted. Up to five people, at time of writing, can collaborate in a voice conference call.

Calls between Skype clients anywhere in the world are free. In theory, you could set-up a videoconference session from London with a friend in Los Angeles and leave it open continuously—a virtual window. Since Skype provides neither the machine cycles to run the two clients, nor the bandwidth between the two machines, it costs Skype nothing.

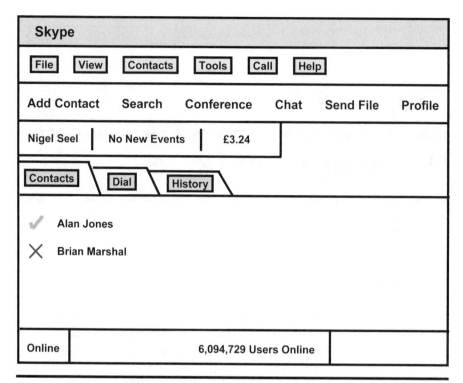

Figure 9.1   A schematic representation of the Skype client.

How Skype Works

Everyone will tell you that Skype's essential differentiator from other VoIP providers is that it is peer-to-peer (P2P). We shall have a lot more to say about P2P in chapter 11, but note for now that unlike most VoIP service providers, Skype does not use its own servers to provide the bulk of its services—users host the Skype service on their own machines via the Skype client. To be quite precise, the key difference between Skype and more conventional VoIP providers lies in the signaling plane. Most everyone else uses protocols like SIP or H.323 that require network servers to maintain directories, locate users, and forward the signaling messages that set-up and tear-down calls. Skype, however, relies on users' PCs and Skype handsets to do these functions.

However, once a call session has been established, all VoIP providers, in the normal course of events, allow the two user terminals to communicate directly across an IP network, without the intrusion of intermediate servers. There are, of course, always exceptions: bridging for conference calls, dealing with NAT and firewall devices, the needs of lawful interception. But in the main, for *all* VoIP systems, media transport (i.e., the call itself) is peer-to-peer.

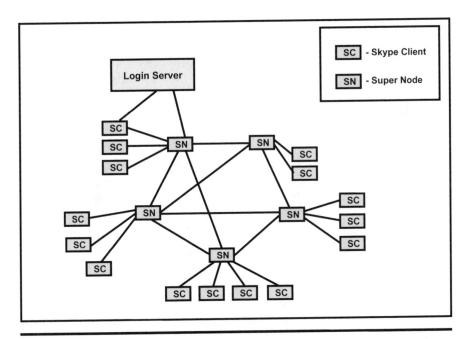

Figure 9.2 The Skype P2P architecture.

As shown in Figure 9.2, there are two kinds of Skype nodes, an ordinary Skype client and a super node. Skype clients are the programs used by people who want to make and receive Skype calls. Super nodes are also Skype clients, but running on these machines the client performs extra functions to help make the Skype service work as a whole. These extra super node functions can include:

- Acting as a proxy between a Skype client and the Skype log-in server,
- Helping a Skype client determine if it is behind a NAT/Firewall device,
- Helping Skype clients to find other Skype users,
- Proxying calls between Skype clients behind NAT devices and/or firewalls,
- Providing conferencing services (3+ way calling).

There is no direct way for a user to control whether their Skype client program becomes a super node or not, but promotion to super node does require that the machine running the client should have a public IP address and should not be behind a NAT device. Machines with more powerful CPUs, more memory, and more network bandwidth are also preferred. Many home machines will, therefore, never become super nodes.

In experiments described by Baset and Schulzrinne (2006) a survey was made of super nodes on the global Internet. The United States had 84 percent, Asia had 9 percent, and Europe had 7 percent of identified super nodes. In more than 8,000 logins, 35 percent had a North American university suffix (.edu) and these comprised 102 universities. The top five universities were, in order, Harvard, Columbia, the University of Nebraska-Lincoln, the University of Pennsylvania, and Boston University.

By allowing Skype clients to undertake super node functions if the machines they are running on have the right properties and capabilities, Skype has arranged that the clients are performing functions which in more conventional architectures are carried out by network servers. It seems clear that many machines have advertised their status as super nodes quite widely, and that the thousands of super nodes constitute a kind of global decentralized directory, describing both each other's existences, and that of connected Skype users.

When you log on to the Skype network, your Windows Skype client will look in a local cache on the C drive to find a super node. It appears to make a random choice, as the local cache lists hundreds of super nodes and their IP addresses. This super node link routes messages to the log-in server (a dedicated Skype server, not a Skype client or super node) that authenticates you. Once logged in, your Skype client asks its connected super node for information about the IP addresses and connectivity status of people on your buddy list. If the super node has this information, it returns it, otherwise it sends back the IP addresses of other super nodes to which the queries should be sent. This process can be executed multiple times. If all else fails, the Skype client can access the log-in server (this is a last option to avoid this server being overwhelmed by traffic). The search process is remarkably fast, with the buddy list often being populated within three to four seconds.

To call someone on the buddy list, the entry is double-clicked. If both caller and callee have public IP addresses, call set-up occurs over a TCP connection set up directly between the caller and callee. If either or both Skype clients are behind NATs and/or firewalls, then TCP signaling is relayed through a super node. Once the call is established, the media transport (the voice call itself) is sent directly between caller and callee using UDP except in the case where both parties are behind firewalls that block UDP traffic. In this case the caller and callee communicate via a relaying super node using TCP with very small packet sizes. Call tear-down is similar to call set-up.

The extra functionality that Skype needs to employ to support calls to and from the public telephone network, the PSTN, cannot use the basic P2P architecture. Interfacing to the PSTN requires special media and signaling gateways, and these are additional pieces of equipment supported by Sktpe as part of its own infrastructure.

The Establishment View of Skype

The setting is a conference in Barcelona on future networking. The speaker, Strato, is from ETSI, the European Telecommunications Standards Institute. He is talking about IMS standards, recalling that IMS is the IP Multimedia Subsystem handling call set-up and tear-down in the official ITU-T/ETSI version of the Next-Generation Network (we discussed this at length in chapter 2). But first Strato needs to talk about Skype.

Skype has been omnipresent at recent conferences. Many speakers admit to using it, and are impressed by its robustness, its voice quality, and, of course, by the fact that it is free to use. For others, it is a major danger to the industry—("how do you compete with free?" asks Strato, rhetorically)—and a veritable road to hell.

Strato's case against Skype rests on five major assumptions:

- ■ Skype cannot guarantee quality,
- ■ Skype is a security risk,
- ■ Skype is not using official standards,
- ■ Skype undermines the carriers' business model,
- ■ Skype is not as good as people think.

Strato speculates that IMS and other parts of the NGN infrastructure will enable *respectable* operators to suppress such rogue traffic, and wonders aloud whether operators would be right to do so.

How worried should we be, I wonder. How *could* operators close Skype down, and would it be in their and the public interest for them to do so? After all, hundreds of millions of people have downloaded the Skype client, and millions are using Skype online at any one time. And there are many, many, other VoIP operators offering free on-net calls. Skype is unique only in using peer-to-peer (P2P) technology (invisible to the user) and being really easy to use.

Skype Cannot Guarantee Quality

Skype uses the public Internet as its transport medium. Unlike traditional carriers, it sees the whole world as its marketplace, and this has been reflected in its rapid growth. Skype transmits voice frequencies between 50 Hz and 8 kHz, twice the bandwidth of ordinary PSTN calls, and the improved quality is quite noticeable over even a low-rate broadband connection.

Skype shows that today's Internet provides sufficient quality of service for voice calls. This comes as a perennial surprise to the advocates of Internet QoS add-ons such as bandwidth managers, QoS marking, Diffserv and RSVP, and all the other attempts to introduce multiple service classes to the Internet. However, on relatively uncongested links, Skype voice works just fine without all that stuff.

In fact as far as the Internet core is concerned, there are good reasons to believe that for a network with a rising traffic load, links will normally be uncongested. When a customer signs up with an ISP, they wish to connect to the whole Internet, not just to other users on that ISP's network. In a competitive market, it pays the ISP both to keep its own network uncongested, and also to ensure that peering links with other ISPs are also uncongested. By adopting this policy it cannot guarantee end-to-end performance (some remote ISP could still introduce congestion, although its immediate neighbors might want to have a word). However, by failing to adopt this policy it will absolutely *guarantee* congestion and thereby make itself uncompetitive. The market equilibrium is that a single class of service (best-effort) Internet as a whole will be uncongested.

With an exponential model of traffic increase, most of the additional traffic comes at the end of any time interval. Assuming the network is never allowed to run in a chronically congested state, this implies that most of the time the network will be fairly empty as shown in Figure 9.3.

There is a caveat, of course, in the context of the net neutrality debate. If carriers manage to create a tiered Internet, with a noncongested QoS tier for which Service Providers such as Google and Yahoo! would pay premium prices, then the best-effort service might be allowed to periodically dip into congestion. As VoIP media transport, being peer-to-peer, does not rely upon a Service Provider being in-the-loop of the connection for the call itself, then Skype users (and all other VoIP-using customers) might have to pay extra to have their calls marked-up and carried as high-priority packets. To be precise, end users might received a tiered bill, where they pay extra for the bulk amount of traffic carried at higher

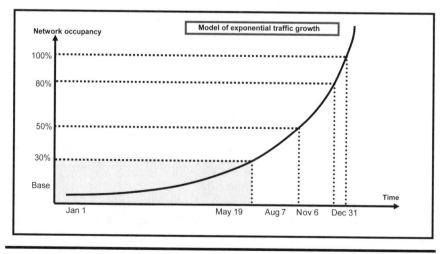

Figure 9.3   Network occupancy without congestion.

QoS classes, this could add an element of usage-based charging, depending on how the tariffing works. It is unlikely that carriers will enforce per-flow billing (too many flows), or be able to specifically identify Skype traffic per se (due to its encryption). Anyway, we are not in that future at the moment, but it is a strong possibility discussed further below. Chapter 13 also discusses these issues in more detail.

The access network is a different issue again. In a home environment, candidates for congestion might include any WiFi link between the Skype client and the home gateway, and then the broadband link itself. Skype voice calls consume from around 24–128 kbps of bandwidth in each direction. But this is well within the upload budget of today's broadband links, and is practically invisible in most users' downlink budgets. Even without class-of-service prioritization, few people today report service difficulties. The next generation of home gateways will support class of service traffic management as VoIP becomes more prevalent, and unless administrative actions are taken to downgrade Skype packets, this should improve the service further in the access part of the network.

Overall, the Internet, left to its natural devices, will not ruin Skype calls. But perhaps the carriers can artificially do it using their new IMS platforms or other packet-inspecting devices? There are several possible approaches.

Introduction of Artificial Impairments

Carriers *could* discriminate in favor of their own VoIP service and against nonapproved VoIP providers as follows. They will arrange for their own clients, running on the PC or authorized terminal, to mark their approved VoIP packets with a real-time service class. Then, at the carrier edge device, they will add impairments such as jitter, delay, and packet loss to any VoIP traffic not to/from the operator's own client, sufficient to disrupt voice quality.

But while this is technically possible, it seems too blunt a technique to work. The carrier will have as customers many other operators and e-commerce sites in the future that will want to support real-time traffic, and not all of them will use the operator's own VoIP system. And the regulator would have a field day in most advanced countries.

Lock-out Skype Infrastructure or Packets

Carriers could close access to the Skype Web site or log-in server via a firewall, or drop Skype packets. There are reports of a few ISPs already having tried this.

In fact, the Skype programming team is engaged in an arms war with ISPs trying to suppress Skype traffic. It used to be possible to stop Skype by blocking the IP address of the default log-in server, according to Baset and Schulzrinne (2006). However, they note that recent versions of Skype get around this by routing log-in requests via intermediate super nodes. In addition, it appears that the Skype executable has hard-coded a number of bootstrap super node IP addresses, which are not made public. Basset and Schulzrinne were able to identify four standardized bytes in the Skype initial log-in message (0x16030100) that could be dropped by a firewall, but they point out that these values are hardly unusual in ordinary data traffic. Most Skype traffic, both control messages and user content, is encrypted, which ensures that there are no regularities that can be filtered.

Could a Carrier Get Away with It?

We already see mobile network operators, as they introduce faster HSDPA data services, banning the use of VoIP. Their reasoning is understandable, but the position seems completely untenable in the longer term. The 3G mobile operators are in a complete bind: to exploit their new networks they have to provide generic high-speed (multimedia) connectivity. While consumers may be pushed into a walled-garden for a while, enterprise staff needs access to their corporate networks and to the public Internet. And with non-usage-based tariffing (preferred by users) the marginal cost of third-party VoIP client usage becomes zero. At this point, their existing and expensive circuit-voice revenues crumble.

And there seems no way out. For every 3G operator who attempts to use administrative means to stop VoIP, there will be a competitor who sees an advantage in permitting it. The average sales person on the road will run Skype (or similar) on their PC regardless of their company's contract with the MNO. Assuming the operator, for technical reasons, can't stop them, will this lead to the threatened contract cancellation? I don't think so.

Skype Is a Security Risk?

The alleged security concern with Skype was one of the more important of Strato's specific issues and a favorite topic of those who wish to attack it. Garfinkel (2005) reviewed the situation with Skype security as follows. Skype claims that conversations, instant messaging and file transfer are all AES-encoded (Advanced Encryption Standard) with 256 bit encryption and that the required symmetric keys are exchanged securely by public key cryptography using a 1024 bit RSA algorithm. Packet analysis confirms that Skype traffic is not readily decipherable. Given the proprietary nature of Skype protocols and algorithms, independent confirmation is not possible, but Garfinkel notes that the PSTN and most competing VoIP systems are not encrypted at all.

As mentioned above, some Skype calls, particularly multi-participant bridged calls, transit super node computers. It is believed that hardly any of these super nodes are owned by Skype—they are mostly machines with public IP addresses owned by ordinary Skype users, frequently university machines as already noted. As such, there may be a security risk for Skype users whose machines end up as super nodes with transit traffic being carried on their machines. There is also the issue of their machine resource being used for supporting other people's calls. Super nodes also seem to be involved in the distributed directory search to set up calls, and when searching for other users logged onto the Skype "cloud."

Skype makes no attempt to hide the identity of connected users. It appears that a Skype search can find any recently logged-on user. This may raise privacy concerns. In particular, organizations may not wish the names and registration details of their employees to be visible to any Skype user in the world. However, it is a Skype license condition of use that Skype is not used for commercial purposes.

Garfinkel's conclusion is that Skype's security status is pretty good for a consumer product, but that there are a number of potential security vulnerabilities and issues, largely due to Skype's unwillingness to be open about how it addresses them. However, it is certainly more secure than most competing VoIP services.

It should be noted that Skype is neither more nor less open about its product than other software companies. Microsoft does not make its source code public, nor does it generally publish its algorithms. There are few independent guarantees that Microsoft security protocols necessarily do exactly what they claim. Most Microsoft users have little idea of what the large number of active Operating System processes on their machines are doing. Users just trust the brand. Now that Skype has been acquired by eBay, one assumes that similar brand guarantees might apply.

It should be relatively easy for a sufficiently concerned organization to test Skype's claims. Its client is under 10 megabytes and could be disassembled. Known plaintext can be sent via Skype's instant messaging subsystem and the cipher text analyzed to assess the power of the encryption. Ditto with voice using a tone generator.

Given the opportunities for competitors to Skype, it is not wholly surprising that Skype see an advantage in retaining commercial secrets. In fact, there is ample room for competitor P2P products that address many of these security issues and that focus more on commercial markets.

Here's the bottom line. Millions of people are actively using Skype at any particular moment. A huge number of vested interests want to discredit Skype on security grounds. Yet not a single case of a Skype security violation has ever been published. And then there is a final twist. Skype was developed by an entrepreneur and a group of Estonian programmers. It was quite likely that for a while, Skype messaging and calling were genuinely difficult to impossible for the world's intelligence agencies to crack. I doubt that anyone would say the same today, following its acquisition.

Skype Is Not Using Official Standards

This was another of Strato's major points, and he felt on stronger moral ground here. Standards are, after all, ethically positive. They prevent lock-in to particular manufacturers, enlarge the market and by increasing competition, lower prices.

Are users locked into Skype? Well, Skype certainly encourages its users to evangelize their friends, even when this turns a SkypeOut call (revenues to Skype) into a free on-net call. These guys are plainly serious about network effects. But it's easy to remove the Skype client and use one of the many other free VoIP soft clients out there. Switching costs and lock-in are minimal.

Would standards enlarge the market in this case? The standards that matter here are for interconnect, not the interior signaling standard. This is not about SIP. At some stage it will be necessary for a Skype user to be able to talk to a Vonage user, for example. However, SkypeOut proves that interconnect is possible, so it is down to Skype to make it happen. Standards and interconnect are not the same thing at all. And lower prices? Well, as Strato himself said, "how do you compete with free?"

It is understandably mortifying for people who have given years of their lives to developing scalable, robust, and sophisticated protocols to find that the most successful product in the world has used something quite different, and proprietary, and secret to boot. But unfortunately, that happens, and it's important not to obsess about it.

It's the reasons for the standards, not the standards themselves, which are essential. In the case of Skype, the issue is not so much the use of a novel P2P protocol as the fact that it's kept secret so that its properties cannot be verified. That in the end may restrict its applicability in higher-value applications.

Skype Undermines the Carriers' Business Model

Skype's cost base is a log-in server cluster, perhaps a few PC bootstrap super nodes, and the salaries and expenses of its few executives and programmers. The operational infrastructure for the millions of Skype users consists of their already existing computers and broadband Internet connections. Skype's fixed costs are low, and variable costs essentially zero. This is how Skype was able to scale to hundreds of millions of downloads and millions of concurrent users so rapidly. Skype makes its money from its value-added services: SkypeOut, SkypeIn, and voicemail.

SkypeOut allows the user to call PSTN numbers from the Skype client on the PC. The call is carried from the Skype client across the Internet for the long-haul part of the call, and then breaks out locally to the PSTN for the final part. Skype has to pay the local telephone operator to carry that final leg, and that charge is pushed back to the Skype user. To use the service, you have to set up a pre-pay account, and Skype retains a proportion of this revenue stream. Per-minutes charges are very low.

SkypeIn allows the user to purchase PSTN numbers in a variety of global locations for around $30 per year (up to ten numbers can be bought). Incoming calls to these numbers are then transferred to the Internet and conveyed to the user's Skype client. This service provides a second revenue stream to Skype.

Skype's third chargeable service, at time of writing, is voicemail, although this is currently bundled at no extra cost with SkypeIn.

From an economics point of view, Skype efficiently utilizes an available resource (user computing power and broadband connectivity) and harnesses it to satisfy a demand at very low cost. This is wholly to be applauded, and is what is meant to happen in competitive markets.

Skype's lack of substantive owned infrastructure means that almost all non-trivial functions must be carried out in the client. If the client is doing infrastructure work for other people (e.g., acting as a directory, transit node, or conference bridge), this may be perceived by the user as effort expended without recompense on behalf of freeloading others. This sets a limit as to how much generic functionality the client can do in pure peer-to-peer model.

The pure P2P model, therefore, works less well where information needs to be managed in a way that is decoupled from any particular user's machine. An example is buddy lists, which cache contact information for family, friends, and contacts. Originally, this information was stored on the local machine but this was irritating if you were running Skype on multiple devices—you had to re-enter the details on each new machine. Now buddy lists are stored on a central Skype server, which adds to Skype's costs.

Another problem for the P2P model is where substantial application functionality is needed. In more conventional architectures, this includes servers for announcements and value-added services. If Skype wishes to introduce these services, it will probably have to put in place special platforms to host them. There are limits to the number of spare university machines!

It is extremely likely that large-scale carrier services based on IMS, and light-weight multimedia-over-IP services based on P2P will both co-exist. There is ample mileage in both approaches. Skype has proved that P2P works and can give excellent service quality, and there are reasons to expect the P2P ecosystem to diversify, perhaps with Skype-like variants for business, provably secure applications, contact centers, e-commerce, and so on. A contrary argument, however, argues for the negative impact of network effects in the absence of interconnect. Why download and install someone else's "secure-pseudo-Skype" when everyone you know is using Skype? Skype's obsession with expanding its user base implies that it could give master classes on the positive network externalities linked with first-mover advantage.

The apparent drawback with a pure P2P architecture is that fixed system functions (directories, conference bridges) are wholly dependent on enough users, and the right kind of users, being logged onto the network. Once this distributed infrastructure is in place, however, incremental end-user facilities are brought by the users themselves. As Chairman Mao once observed, each mouth comes with a pair of hands. A possible hybrid solution for an enterprise or operator is to run a small server farm—a collection of super nodes of last resort—that can also support specialist services. This can be kept up all the time, and ensures that even the second user logging in will get service. Skype may do this itself, but to date, it isn't telling.

Strato, from his standards pedestal, could only see how unfair it was that those cowboy entrepreneurs, refugees really from KazaA, had sidelined his years of hard work on H.323, SIP and now IMS. But we *have* been here many times before. Sometimes the opportunities to be grasped are not those that we predicted with such pedestrian foresight all those years ago. Time to move on.

Is Skype Really that Easy to Use?

Strato might have been on surer ground if he had emphasized usability. I have Skype running on my computers, as do some of my colleagues. The usage patterns are quite diverse. A colleague whose family is in France, but who works in the UK during the week, uses Skype to keep in touch. These are scheduled, laptop-to-PC calls. It seems to work well. Some of my colleagues call me using Skype but it often doesn't work. If I am away from my desk, I usually don't notice a call to the laptop. There are two cases:

- The laptop internal speakers are online, or an external speaker system is plugged in. In this case, I *may* hear the ringing depending on the volume settings, but answering the call is a problem as the headphone/speaker will not be plugged in. Disentangling wires, looking for the right line-out socket and struggling to put the headset on before the caller rings off is not a pretty sight.
- The headphone/speaker is already plugged in—this is rare. However, if I am away from the desk, I will not hear ringing tone as the headphone volume is too quiet.

I am not the only one with this kind of problem. When calling my colleagues using Skype, I encounter similar issues. Is it surprising then that we fall back on the plain old telephony system, with its loud ringing and easy-to-use handsets? Domestic fixed telephony call-costs are low, so unless one is very price-sensitive, there is little incentive to incur the additional hassles of using Skype. And Skype does not substitute for mobile phones except for the smallest niches (where Skype has been installed as a client on a mobile network).

However, things will certainly improve. Even a cursory network search typing the words 'skype handset' into a search engine will turn up an increasing number of USB/WiFi handsets. A WiFi Skype handset can run a full client and does not need a PC or laptop. It seems that if people want to emulate their current fixed phone service with Skype, they will be able to do so. Of course, in marketing, we know very well that just because an existing service can be emulated with a new technology, this doesn't necessarily mean that everyone will defect just like that. Those enticing free Skype-to-Skype calls are mixed with chargeable Skypeout and SkypeIn calls and the handsets themselves are far from cheap. Although voice quality today is good, who knows whether it will stay good—that is outside Skype's control.

Incidentally, not all of Skype's uses are what they seem. The Skype icon features on the system tray, bottom right of a normal Windows XP screen. When the Skype client is in contact with the Skype network, it shows a white tick-mark on a green background. If the client can't find the Skype network, it shows a cross on a grey background. Since Skype is highly adept at making contact, it serves as an excellent indicator of network connectivity. I find myself checking that icon quite a bit.

Conclusions

In economics terms, the Skype service demonstrates an efficient utilization of the relevant factors of production. If the carriers don't like it, they are simply exhibiting the distortions inherent in their current pricing and business models, which positively invite an arbitrage response. In the short-term, of course, it is easier to use administrative action to block innovation than to adapt, but the imperatives of economics cannot be suppressed for ever. In the end, in a competitive market, services that reside on users' already-bought end-systems and exploit vanilla connectivity in the network will be priced pretty much at the marginal cost of said connectivity. In this case, close to zero.

The idea that carriers will thereby go bankrupt or have their business seriously disrupted, is ludicrous. Their market structure is oligopolistic, they retain control of network assets—where barriers to entry are severe, and all the major fixed and mobile players have significant market power. The only issue is how they will rebalance their tariffs nearer to their real cost structures, and the extent to which they intend to rely upon extracting rents for vanilla services such as network access vs. their abilities to develop new revenue-generating services where they have a competitive advantage (integrated network-hosted services, to be specific). The global herd of fixed and mobile oligopolies will have to lumber to a new business model without too much overt collusion, but lumber they will, and they will get there in the end.

In conclusion, at the moment, Skype still looks niche, but service innovation in the P2P space can be fast, and if they can put together something new with real mass appeal, then they could seriously frighten the IMS-bound carriers once again. But don't forget, the carriers control the infrastructure.

References

Baset, S. A., and Schulzrinne, H. G. 2006. An analysis of the Skype peer-to-peer internet telephony protocol., Technical paper. New York: Columbia University.
http://www.eecs.harvard.edu/~mema/courses/cs264/papers/skype-infocom2006.pdf.
Garfinkel, S. L. 2005. VoIP and Skype security, Technical Paper. March. http://www.tacticaltech.org/files/Skype_Security.pdf.

Chapter 10

Spectrum Auctions

Introduction

It was a sunny summer morning as we sat in a conference room high above the Rue de Rivoli in central Paris. All the windows had been flung open, and I could see the Jardin des Tuileries to the south. Distant murmurs of cars and tourist chatter filtered up from street level, all suggesting it was far too pleasant a day to be indoors. But I had no choice. Having arrived by Eurostar from London yesterday evening, I was now deep in discussions with my French colleague, Dromio, on a proposal we were putting together. The European Union frequently announces research frameworks, and invites bids for projects. We were bidding for one that would assess likely developments in fixed-mobile convergence over the coming years and that would propose options for EU regulatory policy. One of the critical issues the tender had highlighted was the procedure for spectrum allocation, and we were invited to state our opinions. Dromio's view was plain and incisive. "The cost of the 3G licenses crippled the entire mobile industry. This must never happen again! We must stress in our submission that other, fairer methods of spectrum allocation must be used in future."

As the hairs on the back of my neck began to bristle, I recalled that Dromio used to work for a major European equipment vendor, which had undoubtedly seen a serious cutback in orders from the debt-laden mobile sector subsequent to the Internet boom. But nevertheless, I was sure he was wrong. The 3G auctions, and auctions in general, were not a bad thing but a good thing. While the vast sums bid for the mobile licenses might or might not have had an adverse effect on the industry, my instincts were that from a public interest standpoint, they actually constituted a positive outcome. And specifically, I did not believe they had raised prices for the public.

The UK Auction of 3G Spectrum

The debate about 3G spectrum auctions centers around the UK case, which started the European round. In the spring of 2000 the British government auctioned five 3G mobile licenses on 20-year leases. After 150 rounds of bidding over seven weeks, the auction had raised £22.5 billion ($34 billion) for the government and the licenses were in the hands of Orange (now France Telecom), Vodafone, BT (now O2, bought by Telefonica), 121 (now T-Mobile), and TIW (now 3). The winning bids ranged from £4 billion to almost £6 billion in the case of Vodafone. Similar auctions were subsequently held across Europe with mixed results.

Almost immediate, however, there was a backlash. A unanimous opinion was that the Mobile Network Operators (MNOs) had overpaid, that the result would be higher prices to consumers and a delayed 3G roll-out. It was demanded that the government should refund some of the money back to the industry (it refused). Economists were more laid back (see Kay 2004; Cramton 2001), claiming that the licenses were a sunk cost having no relevance to forward pricing decisions. The worst that could happen was that if shareholders believed their companies had overpaid, then the lack of return on this investment would lower the share price (as the discounted sum of future earnings). I would add that there was an additional risk that the regulator would go easy with the MNOs to compensate for their debt-servicing charges and to help restore liquidity to the industry—a back-door subsidy,

Where does the truth lie? Many industry players continue to maintain that auctions place too heavy a burden on the industry, weakening operating companies and their suppliers, and harming the public. With new spectrum blocks becoming available on a regular basis, getting the allocation mechanism right is an important factor in smoothing the way to Next-Generation Networks. To understand the impact of a significant spectrum costs to MNOs, we first need to look at how pricing works in the telecoms business.

Pricing in the Telecoms Business

The telecoms business is very far from being a perfectly competitive market. Until recently, it was assumed to be a natural monopoly due to the enormous fixed costs involved in building out a national network, and the consequent increasing returns to scale as more and more customers paid to use such an expensive network at very small marginal cost.

In most countries there are only a small number of facilities-based MNOs. The high fixed costs of a network have to be spread over a sufficiently large number of paying customers. As the number of operators increases, each gets a smaller and smaller slice of the market cake, and at some point the business case for the subsequent market entrant collapses. Mobile *Virtual* Network Operators (MV-NOs) are not so restricted, as they resell capacity on an already-existing mobile network infrastructure.

Four to six players, which is what we normally see, is still too small a number to foster unrestricted competition, although chapter 6 would predict that the long-term market would have only three. Instead, as noted there, we see a form of market called oligopoly. The operators desperately wish to operate as a kind of "collective monopoly" to achieve maximal returns, but as there are market-share rewards for cheating, an air of instability pervades the arrangements. This is accentuated by laws that normally exist to restrict collusion and the formation of any kind of price-fixing cartel.

Cost and Demand Curves

A basic tool for thinking about the pricing options open to a MNO can be seen in Figure 10.1. In Figure 10.1, unit prices and costs are on the vertical axis, and the number of units sold are on the horizontal axis. A unit of service is whatever you are buying per month: call minutes, handset or line rental, and megabytes downloaded.

The short-run marginal cost to the operator is whatever it immediately costs to supply you with that service: perhaps some fraction of the cost of a subsidized handset plus customer service plus fractional operating costs. Note that to a first

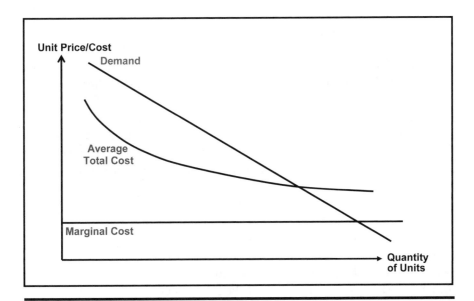

Figure 10.1 Short Run MNO Cost and Demand curves.

approximation, and in the short run, this cost is constant—independent of how many other subscribers are already on the network. The marginal cost is shown by the horizontal line.

The operator's fixed costs (rents, debt repayments, out-payments for leased lines, permanent staff salaries) also have to be covered of course, and from the point of view of business survival, what matters is the total cost (fixed + variable) and how it is averaged over an increasing number of customers units sold. The *average total cost* line on the graph shows how the sum of fixed and variable costs goes down on a per-unit basis as the quantity of units sold increases—more customers are paying for the fixed cost of the network and staff.

If the price is set high, then only a few very keen customers will buy. If the price is very low, assuming the service offered is genuinely popular, then many people will spend their money.

This basic truth is shown by a demand curve sloping from top left to bottom right. When the price is high, few units are sold. When the price is low, many units are sold. So, how should the operator set the price?

The Profit-Maximizing Price

By sliding down the demand curve from left to right, the operator adjusts the price and therefore the number of units sold. Assuming everyone gets to see the same price, the profits after costs of sales is the area of the shaded rectangle (since the marginal cost is taken to be constant):

$$(\text{price} - \text{marginal cost}) * \text{quantity-sold}$$

This is true for *any point* on the demand curve. The gross profit rectangle starts tall and thin, then becomes rather square looking, then becomes short and wide. The area is the maximum somewhere in the middle—shown in Figure 10.2 as the monopoly price. Here the operating profits are highest.

Any sane company would want to sell at the monopoly price, so what stops them? The answer is *competition*, and in its absence, regulation. If you are selling at the monopoly price and I am competing with you, then I only have to set my price a little lower, and customers will desert you and flock to me (other things being equal). Sure, I make less money than if I was able to sell at the monopoly price, but I am still more than covering my operating costs.

It is not difficult to see that in this simple model you get a race to the bottom—a price war. And how low is the bottom? If I am determined to take business away from you, I am prepared to sell an additional unit at *any price down to my short-term marginal cost*—I will still make incremental revenue on the deal. In telecoms, as in many high-tech industries, the cost structure is high fixed-costs

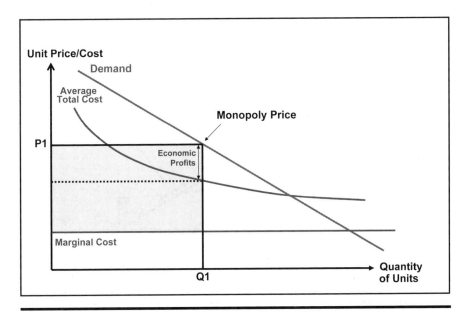

Figure 10.2   Profit-maximizing price.

and low marginal costs. Typically the marginal cost of producing the extra CD, piece of software or delivering one more telephone subscription is very low, and specifically below the average total cost. If you price at short-run marginal cost, then you cannot pay your overheads and you will go bust. But not necessarily straight away.

In fact, there are a number of marginal costs over increasing periods of time. The shortest time period is the extra cost to allow, for example, one more copy of a software product to be downloaded by a customer over the Internet. This cost must be close to zero. Over a longer time period, a marginal cost can include that fraction of costs such as salaries, rents, utility bills that on a shorter time frame are taken as fixed. And on a longer time period, several years, marginal costs include those capital costs involved in growing and modernizing the company's plant and equipment. To stay in business, prices have to be sufficient to cover the latter costs, which is why regulators talk about LRIC as their regulated price target—Long Run Incremental Cost (Figure 10.3). Notice that this is not quite the same as average total costs, since these are backward looking (Kahn 1998). Average (unit) total costs do not necessarily have to be covered by the regulated unit price—it is not the job of the regulator to allow pricing levels that insure against poor business decisions. To stay in business, prices have to be sufficient to cover all current costs and projected forward investment, and that is what LRIC is measuring (and one reason why it is so hard to determine).

However, business is carried out day-to-day, and in a short-run competitive situation my fixed costs are indeed fixed. It still pays me to price down to short

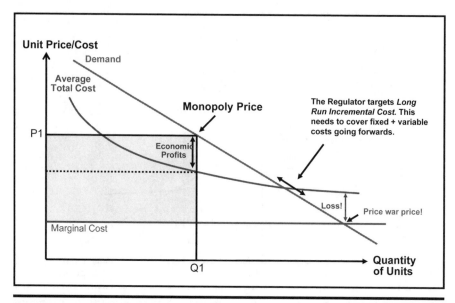

Figure 10.3   Regulated Price.

run marginal cost to take business away from you. Yes, neither of us will recover our fixed costs, and in the end, one or both of us could go out of business, but maybe it will be you. There is, after all, a reason why the most frequent adjective applied to price wars is ruinous.

Strictly speaking, if I have deep pockets, I could price down to zero and definitely bankrupt you if you have less financial resources than I have. This is called predatory pricing and is usually illegal (although it can be hard to prove). My incentive to do it is that I destroy competition and can recoup all my losses later by monopoly pricing. Well, we all think we know this is bad, but what exactly is wrong with monopolistic behavior?

The standard critique of monopoly is that the monopolist under-produces. There are customers who would be delighted to buy the product (e.g., a mobile phone service) at a price that would certainly cover the operator's costs, but the operator declines to serve them at that price. However, monopoly pricing may be the rewards accruing to innovation, prices will come down as the premium returns attract competitors. And sometimes, as in the pharmaceutical industry, apparent monopoly prices are covering the cost base of all the drugs that didn't make through to production. Here marginal costs are far, far, lower than total costs.

As we observed in chapter 6, a level of premium pricing in high-tech industries is essential to give some headroom to the company to innovate. Yes, capital for innovation could always be raised from the capital market, but lacking in-house expertise, the costs would be higher. It's similar to the discussion about the correct limits of the firm, the point where administrative overhead outweighs market transaction costs.

Did the 3G Auction Lead to Higher Prices?

Peter Cramton (2001) spoke for many economists when he argued that the license payment was a one-time fixed fee and that it was a sunk cost that could not affect subsequent pricing decisions (which would be made in the profit-maximizing way as just described). He did suggest that the debt burden on the operators could well inhibit them from acquiring further debt to finance network build-out, and that this might inhibit service roll-out, or even lead to bankruptcy.

It might be argued that debt also has to be paid for. Regular interest payments (and repayments of the principal) are another fixed cost for a company. Increasing debt pushes up the Average Total Cost curve, and therefore the pricing point where the company makes a normal, competitive rate of return. If you like, it raises the price floor for sustainable business for operators in the sector. This is all true but beside the point. When the auction contenders put in their bids for the spectrum licenses, they were proposing an investment just as if they were considering funding any other capital program. All investments by definition incur costs, and the usual criterion as to whether one should make them is positive NPV (net present value). To make an investment rationally, you must believe that the NPV of the investment is positive, which means you will more than recover your debt-servicing charges.

Each bidder would have put together a financial model, where cost items included 3G license fees, network build-out costs, service development and marketing charges and so on. If the anticipated (discounted) revenues covered the summation of all the (discounted) costs in the model, then it was rational to bid. Every serious bidder had constructed such models, and in reality, few were surprised by the size of the winning bids. This last remark is, of course, tautologous: no one would have bid past a point where they still considered they would (eventually) make a profit, the auction design successfully forced companies to bid (very close) to that limit. No doubt the auction winners would have liked to have bought a portion of their input goods at bargain-basement prices (i.e., via a public subsidy), but it was not to be (at least in the UK). And note that debt servicing costs are fixed costs—they make no contribution to marginal costs, and therefore do not affect the monopoly price point at all (or pricing decisions in general). This was Cramton's point.

However, the mobile phone industry is not heavily regulated as regards consumer price. And as good oligopolists, the MNOs make every effort to avoid a price war with each other. Instead they indulge in endless efforts to develop market segmentation through which they can practice price discrimination, charging more to customers prepared to pay more. So there is little evidence that the debt-driven elevation of the price floor is actually visible in customer pricing—the observed prices based on market strategy are simply higher than that. The oligopoly is simply making less economic profits than it otherwise would have done. Note that if the government had run a less-effective auction (or had capitulated to the demands for a refund), then it would simply have been pricing spectrum at a lower value than the assessment of the operators themselves. This would have constituted a subsidy to the industry.

Maybe the mobile phone sector should be regulated more stringently? Perhaps, but it is not very likely. The market is quite competitive already, particularly since the arrival of Mobile Virtual Network Operators. Since these can get quite low wholesale rates (which could, in principle, even be below long run incremental costs for the facilities-based MNOs) they can compete quite aggressively on price, as we have seen in a number of European countries. Other technologies such as voice over IP over competing radio technologies (WiFi, WiMAX) should also increase price competition in the longer term. A surfeit of regulation should be avoided, it tends to create "rent-seeking behavior"—costly activities aimed at influencing the regulator rather than on increasing value for the customer. More competition is a considerably better motivator for innovation and lower prices.

We should conclude that the 3G bidders paid the correct market price for spectrum, namely what they thought it was worth. If they had paid less, by poor auction design or via a "beauty contest," the government would have been granting them an input factor of production at less than market price. This is called a subsidy. Given the oligopolistic structure of the market, where the players have market power, pricing is not even that of competitive markets (i.e., priced at long-run incremental cost). Even if the government *were* so ill-advised as to return some of the money, there is no reason to believe that prices would come down, although shareholders would be pleased. It is possible to imagine a mobile industry that is heavily subsidized by the government (i.e., the public) and is very competitive or very tightly regulated. Under these conditions, prices could indeed be made very low. But this is not a sustainable economic venture as we understand it. It is preferable to rely upon technological advances, economies of scale and vigorous competition via MVNOs to address pricing issues. This seems to be working its magic as the experience since 2000 bears out. If anything, the regulators have been too soft on the industry, particularly in European roaming, where the price-cost ratio has been particularly wide. Over the last few years, we have seen an increasingly diversified and structured market, with service packages and price points at every level. The industry lobbyists have moved their attention to other matters, and there has been very little debate about the alleged 3G auction cost overhang.

I did not put these arguments to Dromio. We didn't have time, and he would not have been convinced. The issue was one of corporate loyalty to him, not economic logic. What is in the public interest is not always in the interests of operators or their equipment vendors. We put something more anodyne in the proposal and moved on.

Appendix 10.1: Auctions in More Detail

Introduction

Until comparatively recently, radio spectrum was allocated to broadcasters and mobile operators through a form of a beauty contest. Applicants had to submit business plans and were assessed by government committees against numerous criteria such as who would guarantee the lowest consumer prices, the most extensive coverage, stimulate the most creative usage, and so forth. Of course, such plans cannot credibly predict the market conditions some years hence, so such contests were, in reality, opaque public subsidies to the industry, prone to lawsuits from losing contenders, and the ever-present possibility of favoritism and corruption. By contrast, an auction process, if well-designed, can identify the honest valuations that businesses put on assets such as spectrum, and can also raise revenues to pay for government without the distortions and disincentives causes by taxation.

In this appendix, I will first look at the different kinds of auction and discuss optimal bidding strategies. I will then look again at the experiences of mobile phone spectrum auctions in Europe and show how poor design can permit bidder tactics such as collusion and predation that can wreck the auction as a means to develop competitive markets as well as a generator of revenues. Organizations taking part in large public auctions put together high-powered bidding teams to advise them on strategy and tactics. Neither this appendix, nor even its references (e.g., Klemperer 2004), can substitute for those. But understanding something about the principles and pitfalls of the auction process can be valuable to managers in the context of the next-generation network. This is not just about radio spectrum, although there are numerous spectrum auctions planned, such as those for further 3G, the sale of WiMAX frequency band licenses and the disposal of analogue TV spectrum following digital switch-over. With Internet trading, many commodity items—some high-value such as IP routers—are being auctioned. And the standard RFQ purchasing process is also an auction, albeit one where the objective is to buy at the lowest price (other things being equal) rather than sell at the highest. Auctions have been around for a long time, and as prologue, let us consider one of the strangest.

The Auction of the Empire

The Praetorian Guard had been set up by the first emperor, Augustus, in 27 BC as an instrument of his personal power (Gibbon 106–109). Comprising ten cohorts of 1,000 men each, they were stationed in and around Rome. The next emperor, Tiberius, moved the whole Guard to a specially-built citadel within Rome itself, in the early years of the first century. Around the time of Vespasian, mid-first century, they were increased to around 16,000 elite troops.

The Praetorians soon realized their power, not simply to support emperors but also to dispose of them. In AD 193 they had murdered the emperor Pertinax (Wells 1992, 256). The reasons are not entirely clear: Pertinax had been in office for only three months, and had himself been implicated in the murder of the previous emperor, Commodus, who had proved himself singularly ineffectual. However, with Pertinax's demise, there were no obvious successors.

Sulpicianus, the father-in-law of Pertinax and a leading public official, was endeavoring to calm the roman masses after the assassination when the Praetorian Guard marched up bearing his son-in-law's head on a lance. Astonishingly, Sulpicianus attempted at this point to claim the mantle of emperor himself, but the Praetorian leadership, sensing a better deal, ran to a nearby vantage point and proclaimed to the waiting crowd that the empire would be disposed of by public auction.

This offer eventually reached the ears of a wealthy Senator, Didius Julianus who was sumptuously dining at the time. He made his way to the Praetorian camp and began to bid against Sulpicianus from the foot of the ramparts. Sulpicianus had already bid a $25,000 *donative* for each soldier in today's money, when Julianus submitted a *jump bi'* of $32,000 per soldier. The purpose of a jump bid is to intimidate other bidders by indicating that you have a high valuation of the product being auctioned, and thus encouraging them to withdraw early, thus closing the auction and securing a lower price. In any event, the tactic worked. The offer was enough to win the auction and buy the empire.

Note that if all the troopers were to receive this amount, the total bill would have been around half a billion dollars (the rich in Rome were *very* rich). The annual tax revenues of the roman empire at this time were around $7 billion (Duncan-Jones 1998, 37). Given the discretionary revenues available to the emperor, Julianus could have expected to recoup this investment in under a year. Alas, it was not to be. His political support did not extend beyond those he had bribed and three field generals rose against him from opposite corners of the empire. In the end, Septimius Severus, at the head of three eastern legions won, and Julianus was out of office and executed within 66 days, a victim of the *winner's curse*. The Praetorians were also out of luck. Severus ordered them to parade unarmed outside the city, where his Danubian legions disbanded them. Severus subsequently ruled as emperor for the next 18 years.

Four Types of Auction

The Praetorians ran the most common type of Auction, known as the English auction, or ascending bid auction. In this model, the price is successively raised by the auctioneer until only one bidder is left, who wins the object at the final price.

In the Dutch auction (used to sell flowers in the Netherlands), the auction starts at a high price that is successively reduced until a bidder commits to buy. The object is then sold to that bidder at the price they committed at.

In a first-price sealed-bid auction, the contenders each separately and privately submit their bids to the auctioneer and the highest bidder wins, and pays the price they bid. This method can be used for procurement, where vendors submit price quotes and the lowest offer wins. Note that this is functionally identical to the Dutch auction—imagine each bidder in a first-price sealed-bid auction taking their envelope to the auction room. As the auctioneer counts down, eventually the bidder with the highest valuation in their envelope will bid and the auction will end—the Dutch auction.

In a second-price sealed-bid auction, contenders also separately and privately submit their bids to the auctioneer and the highest bidder wins. However, the winner pays the price bid by the second-highest bidder. This is known as a Vickrey auction. Paying the second-highest amount probably sounds mysterious, here is the reason for it (we first have to take a detour via the English, ascending bid auction model).

Suppose Alice and Bob are each bidding for a telecoms license. Alice's valuation of the worth of the license to her is $5 million and Bob's personal valuation is $6 million. Both these valuations are kept secret, of course, for commercial reasons. What does the term "valuation" mean? Simply that Alice would buy the license at any value up to $5 million, and if she paid less she would consider she had had a bargain. At exactly $5 million and at any price higher we assume she would just walk away. Bob exhibits the same behavior, this time around the figure of $6 million.

We run an English auction and, unsurprisingly, as we pass $4,999,999, Alice stops bidding. Bob puts in a bid of $5m and wins the license. But wait, Bob's valuation was $6 million. So, although Bob would have been prepared to bid up to $6 million, the price he had to pay was set by Alice's valuation of $5 million (the second price).

Notice that if Alice had known Bob's valuation, she might have bid past $5 million, pushing the eventual price to Bob up closer to his own valuation, to the profit of the auctioneer and the detriment of Bob's business case. There is a competitive motivation to do this.

Still, if you were the auctioneer, perhaps you would have preferred the Dutch auction? Surely by starting the bidding at, say, $10 million and reducing slowly, then as soon as you reached a touch below $6 million, Bob would have bid and you would receive essentially $6 million, rather than the $5 million you in fact received in the English case. Not so fast, it is very unlikely that Bob would have bid his full valuation. In fact, if he had clearly understood Alice's valuation, then it would have been sufficient just to bid at $5 million and come in ahead of Alice by a whisker. After all, Alice would not have wished to bid her complete valuation either—she, too, would have been looking for a bargain.

In a competitive bidding situation, neither method gives a clear-cut advantage. In the English case, Alice's knowledge about Bob can force Bob to raise his bids nearer his own valuation. In the Dutch case, Bob's knowledge about Alice can permit him to lower his bid.

Notice that the fourth model we looked at, the second-price sealed-bid, the outcome is essentially the same as in the English auction. The winner pays a price set by the highest loser. This is not an accident. It turns out that the second-price sealed-bid auction and the ascending-bid auction are equivalent under many conditions (Klemperer 2004, 14).

Another important distinction is between private-value auctions, where each bidder has their own, invariant valuation of the object(s) being auctioned, and common-value auctions, where bidders might alter their valuations depending on signals (i.e. observed bids) made by other participants.

One of the advantages of the second-price sealed-bid private-value auction (= the private-value English auction) is that the optimal strategy is for the bidder to bid their true valuation (rather than bluff above or below it).

To see this, take a look at Figure 10.4. Case 1 is on the left, where you bid below your real valuation, and Case 2, on the right, is where you bid above your true valuation. The various 'other-bid-n' indicate the possible final bid from anyone else in the auction. A final bid from someone else could be below your bid, in which you pay what they bid (plus the tiniest increment) and win, or it could be a bid that beats anything you bid, in which case you lose and pay nothing.

Take Case 1 where you decide to underbid: if you can win at "other-bid-1," then that is the price you pay, and it is irrelevant that you were proposing to underbid your own true valuation. If the auction is won by someone else at "other-bid-2," then you lost and you pay nothing. Irrelevant again that your own private valuation was higher than the point where you stopped bidding. The final case is "other-bid-3." Here, someone else won the auction by trumping your under-bid. But you regret it, because really, you would have been prepared to bid higher yourself. So underbidding as a strategy is a poor idea.

Overbidding is equally poor. Take Case 2. The first two instances are the same as before, so consider when the other person dropped out at "other-bid-3." You won at this price, but now you wish you didn't, because it is in excess of your real valuation.

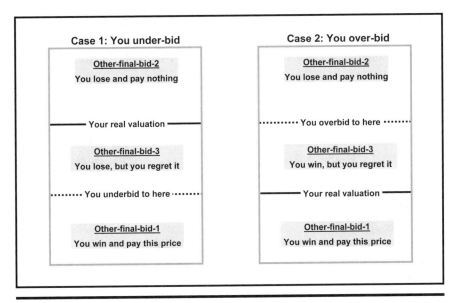

Figure 10.4   Why you should bid at your valuation.

To reiterate, the moral of the story is that in a second-price sealed-bid private-value auction (the private-value English auction), you should bid to the valuation you privately hold.

If we relax the private-value condition, so that each bidder's valuation can be affected by signals they are getting from other bidders as the auction progresses, the equivalence is broken. Now bidders can learn something from the English auction that they cannot in the sealed-bid model. Clearly anyone who eventually wins would have had to have had a higher valuation than everyone else. The question they should ask themselves is whether they were right to do so, in the presence of so much pessimism amongst the drop-outs. If ignoring the weight of opinion represented by the losers is not a sound policy, we talk about the *winner's curse*, the overpayment to win engendered by excessive optimism about the value of the object being auctioned. This was the situation that faced Didius Julianus when he omitted to ask himself whether the Praetorians could actually deliver what they were auctioning.

Auctions in Real Life

Auction theory is about mathematical models and their properties. Auctions in the real world are competitions where real money is at stake and any and all tactics will be employed to win. Critical tactics to "game" auctions include *collusion* between bidders to avoid bidding against each other therefore lowering prices for everyone, and *predatory behavior*, where weaker bidders are frightened off either before or during the auction, thereby clearing the field and closing the auction early.

Predatory Behavior

We have already discussed the UK 3G auction that attracted 13 serious entrants for five licenses. Four of these were the existing UK GSM operators, who were each expected to go after one of the 3G licenses, while the fifth license was reserved for a newcomer. The format of the auction was ascending bid (i.e., an English auction) and after 150 rounds the auction terminated on April 27, 2000 having raised $34 billion. This was around 650 euros per head of the UK population (Klemperer 2004 p. 187).

In July 2000, the auction road show rolled around to the Netherlands. The Dutch authorities decided to adopt the UK's auction design, expecting to net around 10 billion euros or 560 euros per head (opinion was already turning against the prospects for 3G revenues, as the Internet bubble began to burst). But the auction designers missed some significant differences with the UK experience. In the first instance, the Dutch happened to have exactly the same number of licenses to auction as they had incumbent 2G operators, namely five. Existing operators have an enormous advantage over any new entrant as they have operational experience, an existing brand and the opportunity to reuse parts of their GSM infrastructure to lower costs for 3G deployment. It would clearly be difficult for any new entrant to put together a plausible business case that would deliver a valuation above that of the incumbents, or even close to it. That consequent difficulties of displacing an incumbent provided a strong disincentive to participate in the first place. Most of the entrants who had bid independently in the UK auction decided to cut deals with one or other of the incumbents in the Dutch market, with the regulator taking no action to stop this. There was only one further bidder when the auction started, a weak alternate operator called Versatel.

The auction having started, with six bidders for five licenses, one of the incumbents, Telfort, promptly sent a letter to Versatel stating its perception that Versatel must believe that its bids would always be surpassed by stronger players in the auction, and that therefore its participation must be motivated by attempts either to raise its competitors' costs or to force access to their GSM or future 3G networks. Telfort stated that it would hold Versatel legally liable for all damages as a result of this.

This extraordinarily intimidating and predatory behavior put the Dutch government in a dilemma, as to take action against Telfort would end the auction early and reduce revenues to a derisory amount. Versatel buckled under the pressure and withdrew. The result was that the auction was a disaster, raising less than 3 billion euros (170 euros per head). The incumbents were delighted at their bargain (Klemperer 2004,155).

Collusion

In 1999, Germany sold ten blocks of GSM spectrum with a rule that any bid had to exceed a previous bid by 10 percent. The bidders were the four incumbents, but the two weaker players, Viag Interkom and E-Plus soon dropped out. The remaining two bidders were Mannesmann and T-Mobil. In the first round, T-Mobil bid low, and Mannesmann bid DM 18.18 million per MHz on blocks 1–5 and DM 20 million on blocks 6–10.

The significance of the peculiar value of 18.18 is that increased by 10 percent, it comes to 20. T-Mobil deduced that this was a signal that Mannesmann would not mind if T-Mobil bid DM 20 million for blocks 1–5 and let Mannesmann win the remaining blocks at DM 20 million. This result duly occurred and the auction closed at this very low price (Klemperer 2004,105).

Conclusions

Auctions can be an effective method of both creating a competitive market (e.g., by allowing new entrants) and of raising revenues, but neither of these objectives is likely to be in the interests of the proposed bidders. Even a good auction design is subjected to intense lobbying to weaken it, and to permit collusive and predatory behavior on the part of the stronger bidders. If the auction design is sophisticated, lobbyists can often force apparently technical changes past nonspecialist decision makers with the result that the auction fails spectacularly (from the public interest point of view—the successful bidders are only too delighted). Paul Klemperer (2004) cites a number of notable examples in [4].

Knowing your way around auctions, both in theory and practice, is part of the competency set underpinning business strategies for NGNs.

References

[1] Kay, J. 2004. *The truth about markets*. New York: Penguin.

[2] Cramton, P. 2001. Lessons learned from the UK 3G spectrum auction. Economics Department, University of Maryland, May 2001. http://ideas.repec.org/p/pcc/pccumd/01nao.html.

[3] Kahn, A. E. 1988. *The economics of regulation, principles and institutions*. Cambridge, MA: MIT Press, 1988.

[4] Klemperer, P. 2004. *Auctions: Theory and practice*. Princeton, NJ: Princeton University Press.

[5] Gibbon, E., 1776–1788. *The history of the decline and fall of the Roman empire*, 1st ed. London: Strahan and Cadell, 106–109. http://www.everything2.com/index.pl?node=Sale Of The Empire To Didius Julianus

[6] Wells, C.1992. *The Roman empire* (2nd ed.). London: Fontana Press.

[7] Duncan-Jones, R. 1998. *Money and government in the Roman empire*. Cambridge: Cambridge University Press, 1998, http://www.personal.kent.edu/~bkharvey/roman/sources/economy. htm.

Recommended Reading

Paul Klemperer was the principal auction theorist advising the UK government on the design of the 3G spectrum auction in 2000, which raised $34 billion. This was the biggest auction in history, comfortable beating the prior auction of the roman empire. In *Auctions: Theory and Practice* (2004) he reviews auction design both theoretically and in practice—the two perspectives turn out to be very different. This book succeeds in creating in noneconomist readers a sense that they understand the basic terrain of auctions—what they are about—although there is clearly a much deeper set of theoretical results underpinning this map of the territory.

Chapter 11

The Trial of Rete Populi

Introduction

Peer-to-peer is yet another technology trend that swirls around the edges of our IT roadmaps. Computers come with powerful processors, plenty of memory, capacious disk space, and fast network connections. Almost all of this capability is underutilized. Why not leave the computers on all the time, and let them do useful work.

CIOs worry about security and availability. What if something important needs to be done, and the computer with key data on it is turned off by its unsuspecting owner, packed into a laptop bag and then walked out of the door? Far better to spend the money and be reassured by the 7/24 presence of a server farm in a data center.

But there is more to P2P than that, as this modern fable illustrates.

The Trial of Rete Populi

I presume this is some kind of tribunal. I was brusquely told to clean up, given coveralls, and then led to this place. It's something between a court and an interrogation room. There's a judge on a platform to my left and a lawyer at a desk to my right. Across the room three assessors in drab uniforms sit at a long table in bored inattention. I stand in a focus of light, while everything else seems very dark. Truthfully, I am scared as hell.

"Dr. Richard Campbell, you are the inventor and prime developer of Rete Populi?"

"Reh-teh pop-you-lee." At least the lawyer pronounced it correctly. "Rete Populi"—The People's Network" in Latin—has been my consuming project since my doctoral thesis, so this question is confirmation of why I am here, but not an explanation.

"Who are you? What am I doing here?"

"Dr. Campbell, you are not a stupid person, you must know the serious implications of your work. You have provided ample support to criminals, terrorists, and pornographers—did you seriously believe you would be allowed to continue?" A pause. "Answer, please."

Who are these people? They don't look Western, Asian, or African, which kind of lets in Mediterranean, Middle-Eastern, or Latin American. I have no rational expectation that logic can get me out of this, but I try anyway.

"Rete Populi, RP, is a tool for freedom, that's why we developed it. Every previous Internet communication tool has always been insecure. E-mail, Web sites, file-sharing systems like Napster and Gnutella; you can always find out who the sender was, by bullying the ISP if necessary. RP is one of the few systems people can use safely without the government breathing down their necks."

"Very laudable and very libertarian. And yet it still does not bother you that many, shall we say 'less nice' people also have an interest in keeping things quiet from the authorities? Still, we will get to that. Perhaps you could describe how Rete Populi—RP as you call it—actually works?" With a gesture, he indicates the panel behind him.

I am irritated to discover a kind of pride in my creation—how easily we intellectuals are seduced!

"Rete Populi is really Freenet with some additional work to make it easier and faster to use. Suppose I live in a police state and want to distribute a manifesto. I submit it to the RP client program running on my computer, which encrypts it and then distributes the cipher text randomly to other RP nodes. My document is identified only by an automatically-generated Globally Unique Identifier (GUID). I then anonymously post the name of my manifesto, its GUID and decryption key, and my pseudonym anonymously to a friendly, out-of-country Web site (or e-mail the details to my colleagues). If someone wants to get hold of my manifesto, they simply log into the RP network using our client program, submit the GUID, and then the RP network searches across nodes until it finds a copy somewhere. The copy is then moved node-by-node back across the network, and sometimes further cached en route, until they get it. They then decrypt the file using the published key and there you are."

"How could the authorities find the original machine which submitted this document?"

"It's not possible without monitoring all Internet traffic all the time from every machine. Once a document is published to RP, it spreads across numerous RP nodes without any geographical constraints. The nodes do not keep track of the history of document movements."

"The authorities could make it illegal to use Rete Populi clients."

"Authoritarian regimes make many things illegal. A dissident, by definition, is likely to find themselves criminalized. The issue is not whether it is illegal to use RP, but whether by using it anyway, you open the possibilities of exposing yourself or other dissidents through network analysis. We fully expect the secret police to operate RP clients and to use RP to retrieve documents. It is designed that even while doing this, the secret police are still no wiser as to who originally published the document, and they cannot prevent other people also getting hold of it."

"So, we have established that in your view, using your system, people can send and receive any material they like across the Internet and the authorities are powerless to know what is being sent, to identify the perpetrators, or to stop them doing it."

"Two out of three. It *is* possible to know what is being sent if the document description can be obtained: that is, the name, GUID, decryption key, and author-pseudonym. After all, that's how the users of RP are able to find documents and decrypt them in the first place. The police are just another group of users. Obviously, if the descriptions are privately shared, for example by e-mail, rather than being posted on a public Web site, then the police will not necessarily be aware of the material, or be able to decrypt it."

"You are aware of the year 2000 study of your sponsoring network, Freenet, which showed that around three quarters of the material was pornography, material in copyright violation, and drug related?"

"I could sound flippant, and say that these people tend to be the early adopters. In 2000, Freenet had only been up for a year or so. It takes time for news to filter through to dissidents in repressive regimes. In any case, these countries are often poorer and have less Internet presence. The main point is this, though. If you believe that freedom activists should be able to communicate without being exposed to the authorities, then you need a system as powerful as Rete Populi, or Freenet. Anything else is just too vulnerable. Of course, other people with an interest in keeping things quiet from the authorities will use the system, it is a tool which can be used for good or bad ends. The precedents are usually to allow such tools to exist."

"You are, of course, aware that most communication tools, such as the telephone network, and the Internet itself, operate within a legal framework allowing lawful interception of criminal traffic, something which RP conspicuously and flagrantly violates. Let us suppose, for the sake of argument, that we overlook the pornography—perhaps the police will find decrypted files on end-user hard drives to secure convictions, and suppose we ignore the copyright violations—perhaps the record companies' business model will change. Perhaps the real reason we are here is that one person's freedom fighter is another person's terrorist. Would it surprise you to know that major users of RP are jihadists, who coordinate bombing missions?"

"You know the answer. These jihadists also use mobile phones, e-mail, the postal service, white vans, and hotels. All these tools can be used for good or ill, but we do not suppress them because some people use them for bad purposes."

"Actually, Dr. Campbell, it's a question of balance. If we feel that on balance, the tool is more negative than positive, then it is only natural that people with the authority and power to close it down, *by any means necessary*, might very seriously consider that option. You might reflect upon that as we now retire to consider the matter."

Peer-to-Peer File Sharing

Peer-to-peer file sharing programs were used for mass distribution of files, particularly music tracks in copyright violation, from the very beginning. Intensive action to suppress and close down each generation of program led to a subsequent one that was more stealthy and harder to target. The three systems that best exemplify three generations of P2P technology are Napster, Gnutella, and Freenet, although there were, and are, many other programs out there.

Napster

In the beginning, back in 1999, there was Napster, so named after creator Shawn Fanning's nickname. Fanning had been dissatisfied with existing programs for sharing music files, because they were too hard to use. He designed Napster as a combination of client and central directory. You started by downloading the client and allocating a directory on your machine as "shared": this was where downloaded and uploadable music files would be placed. Your next step was to log onto the central Napster server and type a query for the song title or artist you were interested in. Your client then forwarded the request to the central server, which looked it up in the directory and sent matches back, displayed as a list in a window on your client. If you then selected an item of interest and clicked "download," the machine holding that file was contacted directly, and the file copied across to your machine. Finally you could click on the downloaded MP3 and listen (Figure 11.1). Other people were doing the same to access the music files stored on *your* machine.

Fanning had attained his objective—Napster was so easy to use that by February 2001 Napster had more than 26 million users. In some universities, more than 80 percent of the high speed Internet bandwidth was being used for MP3 file-swapping. Its very success led to its undoing. Napster was closed in July 2001 after legal action by the recording studios, who correctly saw it as a huge mechanism for copyright violation. The legal directive forced Napster to close down its central directories. This did not, of course, remove the pent-up demand

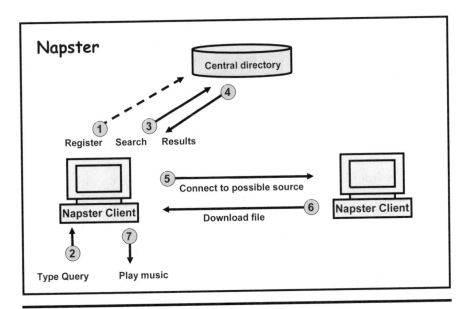

Figure 11.1 Napster.

for Internet file-sharing. Instead, efforts were made to create a more decentralized system that was less vulnerable to legal action. The most prominent of these so-called second generation P2P systems was Gnutella. (Kazaa, currently at version 3.0, is often cited as well, but the authorized client was notoriously infested with ad and spyware).

Gnutella

Gnutella is basically a protocol for joining a network of Gnutella machines, searching in a distributed way among these machines for a file or set of files, and then downloading one. This is how it works.

Initially you need to download a Gnutella client—a popular client is LimeWire (http://www.limewire.com). As part of the configuration of the client, you can choose which directory on your machine will contain material to share (i.e., download and upload). The Gnutella client will typically have hardwired access to an Internet cache of addresses of machines already part of the Gnutella network. Gnutella-connected machines are called "servents" indicating they are both server and client.

To get started you need to find the address of an already-connected Gnutella machine. Details are available from Web sites or various Gnutella servers, described in the client's Internet cache. Once connected to this first machine, your servent sends "ping" messages across the Gnutella network to announce your presence.

Typically, three or four other Gnutella servents may respond with "pong" reply messages if they wish to form a connection with you. It is now possible, just as in Napster, to type a query for a song title, artist or file you are interested in. Your query is sent in the first instance to your immediately connected machines, and they both search their own shared directories for the file and forward the query to machines *they* are connected to. Usually each Gnutella machine connects to three or four others, and queries are typically propagated seven deep. This flooding mechanism means your query could eventually be received by around 10,000 machines.

As you wait for a minute or two, your screen begins to fill up with matching files from the contacted machines. As in Napster, you select one, and that machine is contacted to copy the file to your machine. Wait for it to arrive and it's yours to play, view or whatever as shown in Figure 11.2.

As compared with Napster, Gnutella is resilient against court orders, as there is no central server to shut down. However, there is no secrecy as regards providers of files—to get a file you connect to a machine holding the file, via its IP address. This uniquely identifies the machine and is completely public. However, as hundreds of millions of Gnutella clients have been downloaded, legally going after Gnutella users one at a time does not seem to scale as a legal strategy. This has not stopped the Recording Industry Association of America (RIAA) from trying it anyway, and thousands of prosecutions have taken place in North America. From a performance point of view, about a quarter of the traffic between Gnutella servents consists of query requests, via the massive amplification that occurs through the flooding mechanism. A typical Gnutella query is around 560 bits in length, and with three connections between your machine and its neighbors, a flooded query

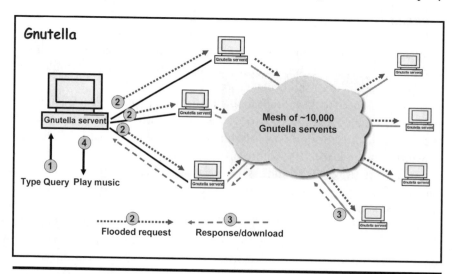

Figure 11.2   Gnutella.

will occupy around 1,680 bits. If you can see 10,000 nodes, then they can see you, and if 1 percent of the nodes launched a query in a given second, the 100 queries per second hitting your link would use a bandwidth of 168 kbps—even these days, that is quite significant, without worrying about the processor load in responding to them: flooding as a routing mechanism is generally considered robust but highly inefficient.

BitTorrent

Gnutella is a "pull" architecture. It relies on people knowing what they want and searching for it. Another well-known P2P system, BitTorrent (http://www. bittorrent.com/) is optimized for publishing material. As well as the usual fare of copyrighted and "adult" material, BitTorrent has also been used to legally distribute large files such as software distributions, game updates, and legal video. Here is how it works.

As usual, you start by downloading a BitTorrent client—there are a number of options. As someone with a file to distribute, you submit the file to your BitTorrent client, which divides the file into blocks of around 250 kB each. It then creates a .torrent file containing the file name, file size, and authentication information about each block, plus the address of a tracker server. Tracker servers hold a tracker directory that will maintain information about where your file blocks will end up as they spread across multiple machines; they also track which machines are downloading which blocks at any particular time. You are now ready to post the .torrent file to a suitable Web site or mail it to potential recipients, which is the stimulus for people "out there" to download your file.

To access a file on BitTorrent, you start by downloading the .torrent file from a Web site (or it might have arrived as an e-mail attachment). You then connect to the tracker server and directory named in the .torrent file to find out who has (blocks of) the file and you then contact those machines to start downloading. As *you* receive blocks, these are stored in your shared directory and can be accessed by other people also trying to get the file. Over a period, starting from the original publisher, blocks tend to distribute themselves across the BitTorrent network, with later downloaders getting the benefit of earlier downloaders' caches. Popular files get disseminated more widely, which makes more copies available for further downloading, which gives BitTorrent excellent scaling characteristics (Figure 11.3).

The collection of machines at any one time involved in sending/receiving file blocks (a torrent) for a particular file is called a swarm. It is considered good practice, once you have received a file, to keep your BitTorrent client active so you can support other users who can continue to download from your machine—this is really the whole point of BitTorrent.

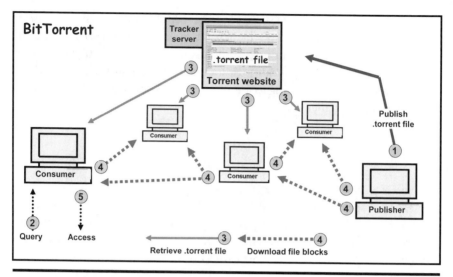

Figure 11.3 BitTorrent.

BitTorrent is extremely effective as a scalable file-distribution system, but it is *useless* for privacy. The .torrent file has publicly available information on the machines providing file blocks, and the tracker directories are a point of legal attack: there *have* been lawsuits. In response, variants of BitTorrent have been proposed that are tracker-less. However, if anonymity and resistance to legal attack is the objective, the current state of the art is Freenet.

Freenet

The details of Freenet operation are rather complex, but the general principles are not hard to understand. In what follows I describe a simple "model Freenet" that illustrates the method of operation. For further details see Taylor 2005, Clarke 2002, and the Freenet Web site (http://www.freenet.sourceforge.net).

Freenet is designed from the ground up to permit the anonymous dissemination and receipt of files in the presence of authorities trying to stop it. All files are encrypted on entry to the Freenet system (large files may be fragmented into smaller blocks) and are passed from machine to machine in encrypted form. Files may also be digitally signed so that, while you (and the secret police) may not know who authored a batch of files, you can both be assured that it was the same individual or group and not an impostor—this permits reputations to be developed.

A file, or document, needs to be named with a descriptive string (e.g., "The-Answer" for human access purposes, and is uniquely identified via the GUID (Globally Unique Identifier) key. In practice, this is a long binary number cre-

Table 11.1   Advertising a Freenet File

Document name	Decryption Key	Author	GUID
The Answer	mK0$6aB	wtns	42
…	…	…	…

ated by hashing the file's description and some extra key information, but in the example below we will just use the key 42.

To advertise a file the process is rather similar to BitTorrent: you need to post a document description on a Web site or send it as an e-mail attachment. The information you provide is shown in Table 11.1.

To use Freenet you start by downloading a Freenet client—there are many to choose from. The initial connection to Freenet, just like with Gnutella, requires finding the IP addresses of some machines already linked into Freenet, e.g., via some friendly Web site. This done, your machine will be assigned an identifier—let's suppose you are node A in Figure 11.4. Unlike Gnutella, where queries are flooded, things are more selective in Freenet. Each Freenet node (i.e., connected machine) has a routing table that contains the GUIDs of files it has on its own hard drive, and also lists at least some of the GUIDs of files resident on other machines it knows about.

Let's suppose you are particularly keen to find "The-Answer," which according to its Web site entry, has GUID key = 42. Your Freenet client first looks to see

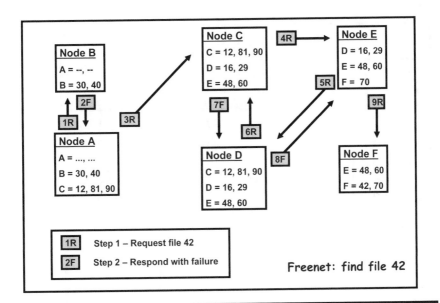

Figure 11.4   Freenet: Finding a file.

if your machine already has a file with GUID = 42. Assuming you don't have it, your client then looks in its routing table to find a connected node that either has it, or that has a GUID that is pretty close. In Figure 11.4, node A sees that the nearest match to 42 is node B, which has a file with GUID = 40, and so A sends a request. It turns out that unfortunately B does not have it and has no further connections to try, B sends back a failure notification. A then tries its second alternative, node C.

Node C does not have file 42 itself, but the nearest match is node E, it forwards the request. E doesn't have it, so forwards it to D, which also doesn't have it, so forwards it to C (having no other sensible choices). Node C has already seen this request, so sends a failure response to D, which having no other choices also sends a failure back to node E. E now tries its other option, node F, and gets a result. Node F does indeed have a file with GUID = 42.

We now move into retrieval mode (Figure 11.5). The file (recall it is encrypted) is handed back to the requestor a node at a time. Each intermediate node can choose to cache the file locally if it wishes, becoming a new source. It can also mark itself as the source of the file, even if it doesn't cache the file locally—its own routing table will point "upstream" to a node that *does* have it. These are all steps to hide the real source from opponents. In Figure 6, node F sends file 42 to node E, which caches it and forwards a copy to node C, which in this example marks E (rather than F) as the file holder, and passes the encrypted file to the original requestor, node A.

Note that node C has no idea whether A wants the file itself, or is merely an-

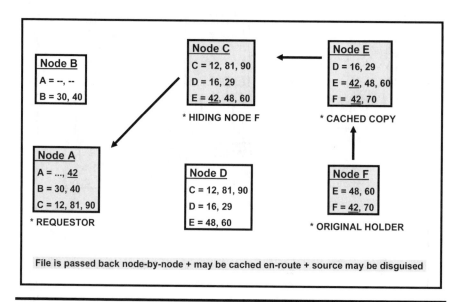

Figure 11.5   Freenet: Retrieving a file.

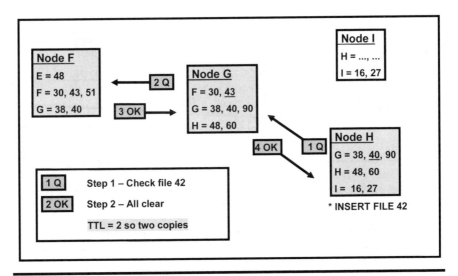

Figure 11.6   Freenet: Inserting file 42.

other node upstream of the real requestor. This indirection removes the situation in Napster and Gnutella where the requestor and provider are in direct and open communication to transfer a file.

You now have the encrypted version of the file you wanted, and by accessing the decryption key from the Web site entry (recall mK0$6aB from Table 11.1) you can decrypt the file, verifying it is from the author claimed, and enjoy "The-Answer." You have no idea who originated the file, and the author doesn't know it is you who has downloaded it. Intermediate nodes are equally ignorant about the ultimate source and destination of "The-Answer," even if they are controlled by the secret police.

Inserting files into Freenet is also an indirect process. You should not assume that node F is the machine of the author of "The-Answer"—what actually happened was this (Figure 11.6).

The author was really using the machine that is node H. He wanted two copies of his document to initially seed Freenet, so he sent out a special query-insert message to the Freenet node his machine knew about with key closest to 42. This message had a time-to-live (TTL) of 2. Since node G has the closest existing GUID of 40, G got sent the message. Node G then decremented the TTL by 1 and forwarded the query-insert message to F as having again the nearest key to 42. F received the message, and decrementing the query-insert message TTL to zero, did not forward it further. F and G finally respond with OK messages back to author machine H.

Node H now sends the encrypted version of "The-Answer" to G, which forwards it on to node F (Figure 11.7). Both G and F cache the file locally and update their routing tables, as does H. In this example, H does not store an in-

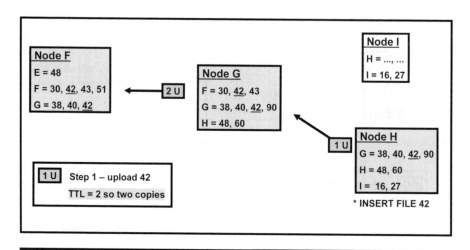

Figure 11.7 Freenet: Uploading the file.

criminating copy of his document locally. And that is how node F came to have "The-Answer" in its local cache when you requested it.

You should be aware that the model just described relates to current Freenet practice. The Freenet development community has been working for a while on Next-Generation Routing, where nodes collect more elaborate performance statistics about their neighbors and try to be smarter about which other nodes should be tried first when searching for, or inserting files. More details on the Freenet site.

Freenet is generally felt to be clunky and slow, the downside of its step-by-step routing and complex decryption processes. Some broadband users have compared it to being back on dial-up, the price you pay for state-of-the-art anonymity. Making a legal case against the owner of an individual Freenet node is claimed to be difficult since they cannot be definitively shown to be either the author or recipient of any of the material in the Freenet shared directory. Since all material is encrypted, the owner can disclaim responsibility for content on his machine. It is quite difficult, in fact, to find out what is on the Freenet partition of your hard drive. To decrypt the files you would have to find a Web site that listed the file with that GUID and its decryption key, assuming they were published at all (GUIDs and keys can also be distributed by other means such as e-mail). Possible sometimes, therefore, but laborious.

The Content on Freenet

What kind of content would you expect to find on file-sharing networks such as Gnutella, BitTorrent, and Freenet? Anyone can do a search on Gnutella content,

Table 11.2   Freenet Contents in 2000

Category	Topic	%
Text	Drugs	59.4%
	Classical texts	8.9%
Audio	Rock	71.5%
	Musicals	9.4%
Images	"Adult"	89%
	Humor	2.6%
Video	"Adult"	76.9%
	Anime, humor, movies	5.1%
Software/ Games	Various	

which will rapidly show a large number of copyrighted MP3 albums available for download. In 2000, Jon Orwant, the CTO of O'Reilly Media, Inc. reviewed more than 1,000 thousand items on the Freenet version current at that time and classified the contents as as shown in Table 11.2

Orwant (2000) commented "if we were to indulge ourselves and construct a demographic of the average Freenet user from Freenet content alone, he'd be a crypto-anarchist Perl hacker with a taste for the classics of literature, political screeds, 1980s pop music, Adobe software, and lots of porn."

To be fair, we do not have an authoritative *contemporary* analysis of Freenet content, although given the likely relative proportions in the world of pornographers, illegal file-sharers and underground dissidents, it seems unwise to get one's hopes up.

In response to such criticisms, the founders of Freenet prefer to talk about empowering Chinese dissidents, and when pushed they suggest that anonymous electronic communications is just a tool, like the telephone network or e-mail—its good uses outweigh its bad uses. In any case, they say, Freenet is not an optimal vehicle for criminal communication as it publishes to the world rather than person-to-person. There are easier ways for small groups of people to communicate in secret (Clarke et al. 2002).

The world of advanced computing is often pretty libertarian, with support for free use of cryptography and against arbitrary control of software and content. Nevertheless, many people feel uneasy about participating in networks where some unknown amount of material on their hard drive is likely to be pornographic or in breach of copyright, even if the legal consequence—due to the design of the network—are likely to be small today. Somehow it seems a high price to pay for providing a secure platform for the dissidents of the world with right on their side.

A Rocky Future for Digital Rights Management?

The massive take-up of Internet file sharing pointed to a large demand for Internet distribution of content (obviously at low cost). What was needed was a way to control distribution so that every act of distribution could also be associated with an appropriate commercial transaction. To put it crudely, you can copy but you'll have to pay. The technical architecture and implementation required to make digital content distribution safe for commerce is called Digital Rights Management (DRM).

The purpose of DRM is to underpin an effective business model for selling digital content. The design and architecture of DRM systems is an attempt to make explicit the legal concept of "rights"—the rules that determine the usage of content through and subsequent to acquisition. People tend to think that DRM is about paying for what you get, and not allowing further free distribution. But rights models are significantly more sophisticated than that. Here, for example, is the list of rights management facilities obtainable from a commercially available DRM package.

- Control of Purchase (pay for a copy before use).
- Forward Lock (no forwarding to anyone else).
- Secure and controlled superdistribution (allows the first purchaser to share content with others. Subsequent sharees must obtain a rights object from the original distributor to unlock the content, and this is chargeable).
- Rental for a limited number of uses.
- Rental for a fixed period of time or to an expiry date.
- Free preview then option to buy.
- Tracking (distributor can register who accesses content).
- Rating (content can be restricted from certain classes of user, e.g., film rating).

Acquired rights determine, via the DRM system, *when* content can be consumed, on what *platforms* content can be presented, how *often* content can be accessed, whether content can be *printed* or *transcoded* to other media, *copied* to third parties, *sold* on, and so forth. Policies, in regard to transfer of some or all of these rights to third parties, are also in scope.

Some people believe that a Digital Rights formal language such as XrML (eXtensible rights Markup Language; see http://www.xrml.org) can completely specify the usage model of a unit of digital content. Others believe this is a grey area, where human (legal) judgment will be forever necessary. A case in point is "fair use," discussed below.

Figure 11.8 details how DRM works. A more detailed walk through can be found in Rosenblatt, Trippe, and Mooney (2004 83–84).

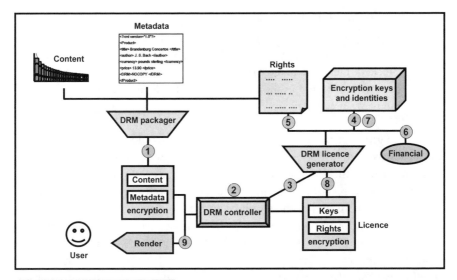

Figure 11.8 How Digital Rights Management works.

- Step 1. The user obtains DRM-protected content, e.g., by downloading from a Web site.
- Step 2. The user tries to access the content. This starts the DRM controller subsystem. The DRM system now tries to obtain a license authorizing it to respond to the user's request. At this stage the user may be prompted for identity information, such as filling in a registration form or entering credit card details.
- Step 3. The DRM controller sends this information to the license generator,
- Step 4. The license generator checks (or inserts) the user's identity in the database and
- Step 5. Establishes the rights the user is entitled to exercise.
- Step 6. The license generator may also bill the user at this point, e.g., through a credit card transaction.
- Step 7. Various encryption mechanisms are enabled, keys generated, and
- Step 8. The encrypted license is sent back to the user.

■ Step 9. The DRM controller now has authorization and means to decrypt the content and play it to the user. The request could have been to print or copy the content, and the rights to do this would have been checked. Device characteristics could have been uploaded to the license generator in Step 3 in the case where the DRM mandates which devices the content is permitted to run on.

There might have been an early belief that DRM could provide a robust and comprehensive support to the market in digital content, and specifically secure distribution over the Internet, but it seemed that as soon as a DRM scheme was brought into service, someone, somewhere would crack it. And once the unprotected content was in the hands of an Internet-connected advocate of the concept that "information wants to be free," it was game over. This situation was formalized in 2002, when Peter Biddle, Paul England, Marcus Peinado, and Bryan Willman from Microsoft wrote a subsequently much-cited paper called "*The Darknet and the Future of Content Distribution.*" In a quasi-mathematical form, they proposed in section 1 of the paper the following three "axioms."

■ Any widely distributed object will be available to a fraction of users in a form that permits copying.
■ Users will copy objects if it is possible and interesting to do so.
■ Users are connected by high-bandwidth channels.

Biddle and colleagues suggested that any copy-protected material will be cracked by someone, somewhere (assumption 1) and that the combination of assumptions 2 and 3 are that the resulting DRM-free material will be circulated widely across the Internet. The network involved in the sharing of DRM-broken material is called the Darknet.

In section 2 of their paper, the authors surveyed the systems we discussed earlier—Napster, Gnutella, Freenet—and focus on the "small world" topology these networks seem to converge upon, which gives them important properties of scalability, performance, and robustness. Sections 3 and 4 covered DRM techniques and watermarking, generally adopting a skeptical tone as to the likely success of these against the Darknet.

In their final section 5, the authors made one of the most interesting points in the document. That the Darknet is a competitor to legal commerce not just because it is free, but also because content without DRM has a significantly higher utility to the user. As anyone in security knows, higher security comes at a cost of greater inconvenience to the end-user. DRM's clunky set of restrictions can be intensely off-putting to consumers, creating uncertainty as to whether the acquired product will continue to function in a number of circumstances where no legal or ethical issues are involved. For example:

- Will a downloaded set of music tracks incorporating DRM still work if copied to my next PC?
- Will I be able to burn the tracks to a CD and, if so, will it play on all CD-players or just those with some special DRM system?
- Will the music play on my preferred soft client or MP3 player?

Who needs these kinds of worries?

As Biddle and colleagues note (2002, section 5.2), "This means that a vendor will probably make more money by selling unprotected objects than protected objects."

The concept of Darknet has been much discussed in recent years. For example, there is a Wikipedia article (http://en.wikipedia.org/wiki/Darknet) that associates the term with the idea of a covert network of people known to each other.

The Greater Utility of DRM-Free Content

Picking up on the Darknet authors' view that there will be a legitimate market for non-DRM-protected material, consider Figure 11.9, a standard demand curve, with a good priced at P selling quantity of units Q. The revenue to the vendor is then P * Q. Note that there is further demand to the right of Q, and that if the price was dropped to zero, then up to Q' units would be taken (at zero revenues, of course). Suppose someone covertly distributes instances of the good to the people in the (Q' - Q) part of the market—either free, or at a price that undercuts the market price, without affecting the Q people who continue to pay P. Note that this is a form of price discrimination. Is this a problem?

Well, it ought not to be. Those people would not have bought the product at its market price anyway, so the fact they now have it has presumably increased social welfare while not diminishing revenues at all. Obviously these hypothetically cheaper or zero-priced goods must not be available in the main market, otherwise they undercut those Q goods sold at P and so destroy revenues. This is the well-known arbitrage problem of price discrimination.

This kind of situation occurs with information goods in third world countries, where piracy is rife. Since the people are poor, they could not afford the market rate in the first place, so would never have bought the product. The cheap copies they *do* buy (or acquire for free) are difficult to re-export to first world countries due to legal disincentives there. Vendors who experience this phenomenon, such as Microsoft, have often seen the positive side of such piracy, as a form of market training and stimulation, almost a loss leader. Once hooked on Microsoft … . As the country gets richer, pressures increase to enforce copyright and to make sure products are sold legitimately at market rates.

Looking again at Figure 11.9, the triangular area marked "Unsatisfied demand"

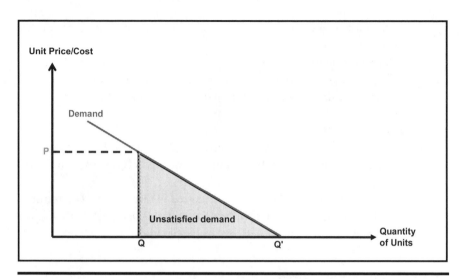

Figure 11.9   Sometimes copying does no economic harm.

can be genuinely closed off, secured from piracy and subsequent arbitrage, with effective DRM. But with today's DRM, piracy is not the only thing stopped dead. Informal copying to friends and family can also be suppressed altogether, possibly including copying from your present self to your future self with a different playback machine. DRM can also easily stifle traditional "fair use" (backup copies, copies for the car, etc). This is one reason why DRM has such a bad name.

One brief caveat: the Darknet thesis still applies—it will always be possible to acquire zero-price digital content if you are prepared to collude in illegality. The various P2P systems described earlier in this chapter make sure of that. However, the market size of people prepared to act beyond the law as receivers of stolen property is limited, so we can consider this as a kind of "background noise" that acts to subtract a bounded amount from the overall demand curve. Fair use and limited copying for oneself, to family and to (genuine) friends is something else, and apparently not painted with the same moral brush. (We know they would not have paid for the content in their own right, right?).

Incidentally, these are the kinds of reasons that "fair use" was so legally controversial in the first place. Content owners were terrified that recording music from the radio or films from TV, for personal or family "fair use" would destroy the market for CDs, video-cassettes, and latterly DVDs.

Now, suppose, as in Figure 11.10, that suppliers emerge prepared to sell content with no DRM. They provide MP3 files rather than WMA with DRM, for example. From the customer's point of view, this is a more attractive product as there is far less hassle and risk. Demand is higher at the same price, and the demand curve shifts up and to the right. We are assuming for this example that the suppliers are not competing on price but on their lack of DRM.

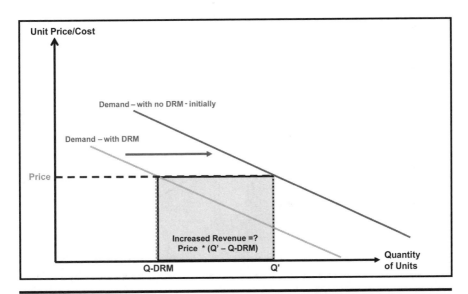

Figure 11.10 Demand for Content without DRM—First phase.

In theory, these enlightened suppliers tap into an enhanced revenue pool of size P * (Q' - Q-DRM). However, one of the reasons that customers are flocking to their online stores are that they can share their purchases with their friends and families (to whatever is the limit of their moral conscience). These grateful beneficiaries of free copies then subtract demand from the market, pushing the no-DRM demand curve leftwards (Figure 11.11).

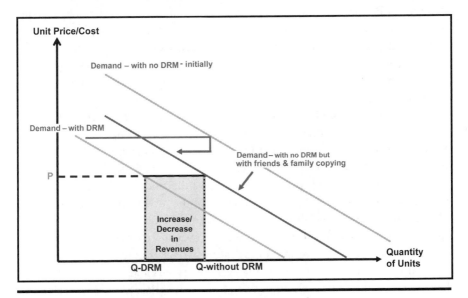

Figure 11.11 Demand for Content without DRM—Steady state.

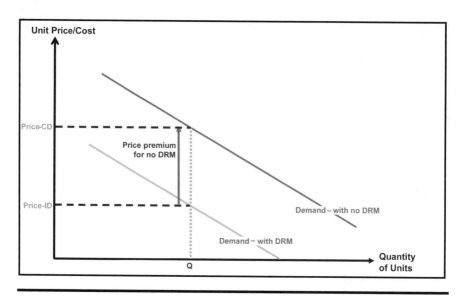

Figure 11.12 Price premium for content without DRM.

Are our good-guy retailers still making money, or has the market contracted due to content sharing to below the original DRM-limited level? How far are DRM-protected and DRM-free products substitutes? It's impossible to say in general, because things like competition, market share, attitude of original content owners, and attitudes to illegality all play a part. However, in case you think the example might be fanciful, I offer my own illustration.

I buy non-copy-protected music CDs and rip them to my MP3 for use on my PCs (and MP3 player) because I don't want to end up with DRM-restricted content. I have no desire whatsoever to facilitate illegal file sharing, I just don't trust DRM packages. In this view I am heeding the advise of most computer journals, which recommend exactly this policy. In fact, I am probably paying a *premium* price for the CDs for this freedom rather than straightforwardly downloading DRM-protected content from Internet music stores, as shown in Figure 11.12.

The above discussion is based on Shapiro and Varian's analysis of DRM (1999, 98–100). The model presented here is a little different to theirs—Shapiro and Varian have a more inelastic curve for non-DRM-protected products than for those with DRM, which simplifies their explanatory analysis.

A final point. We tend to talk about DRM independently from the media it is supposed to be protecting. The examples we use are mostly about DRM applied to music tracks, reflecting contemporary piracy concerns. But music has a special characteristic: people like to listen to it over and over again. This fuels the concerns I mentioned above about the destiny of one's own music collection over a listening-trajectory of perhaps many years. But films and books, for example, have nothing like the same reuse value. Apart from those people who claim to have watched *The Sound of Music* or *Starship Troopers* 20 times, most films on DVD today are not watched more than two or three times. This may make the DRM overhead considerably less onerous.

Summary

In this chapter we looked at a number of peer-to-peer file sharing systems and discussed some of the issues behind their design and use. Originally used almost entirely for illegal sharing of copyright material, mostly music tracks, P2P systems have developed as legitimate content distribution systems and also, more controversially, as anonymous platforms supporting freedom of speech.

Internet distribution of digital content is fast, easy, cheap, and opens up a global market. These attributes are highly attractive to business if their wares can be safe-guarded from piracy. Enter Digital Rights Management. However, DRM systems find it extremely difficult to formalize the distinctions between criminal duplication and fair use. They tend to err on the side of restricting fair use, which infuriates legitimate users who end up not being able, for example, to transfer their expensive music collection to a new machine. This implies there is a market for publication of material without DRM, or with the very lightest of DRM touches. This market demand may be an obstacle to the recording industry's petrified vision of a future of DRM-protected content everywhere.

The jury is still out on the fate of Dr. Richard Campbell and Rete Populi. I venture to predict that while it is small-scale and somewhat difficult to use, Rete Populi will be tolerated. If it gets past a threshold of usage by people powerful states treat as enemies or criminals, then either administrative/legal action will be taken against Dr. Campbell and his colleagues (there are always grounds) or some kind of trapdoor will find its way into the system. The imperatives of *legal intercept* will not be denied.

References

Biddle, P., England, P., Peinado, M., and B. Willman. 2002. The Darknet and the future of content distribution, p. 2. http://crypto.stanford.edu/DRM2002/darknet5.doc.

Clarke, I., S. G. Miller, T. W. Hong, O. Sandberg, and B. Wiley. 2002. Protecting free expression online with Freenet, *IEEE Internet Computing*, 2000 Jan-Feb.. http://freenetproject.org/papers/freenet-ieee.pdf..

Orwant, J. 2000. What's on Freenet, *OpenP2P*. November 21. http://www.openp2p.com/pub/a/p2p/2000/11/21/freenetcontent.html.

Rosenblatt, B., Trippe, B., and Mooney, S. 2002. *Digital rights management—Business and technology*, New York: M&T Books.

Shapiro, C., and Varian, H. R. 1999. *Information rules*. Cambridge, MA: Harvard Business School Press.

Taylor, I. J. 2005. *From P2P to web services and grids.* New York: Springer.

Recommended Reading

Ian Taylor's book *From P2P to web services and grids* covers all the P2P systems discussed here at the next level of detail and is recommended. It is also a source for more details about Grid Computing, which we discussed earlier in chapter 4.

Chapter 12

Machines Who Talk

Introduction

In 1979 Pamela McCorduck published *Machines Who Think*, a survey of the then-nascent field of Artificial Intelligence (it was reissued in 2004 with an Afterword added). Apart from the shock of the sentient pronoun, McCorduck's book also helped raise the hype-level of AI in the 1980s. The field has tended to disappoint—technologies such as expert systems promised much but seemed to vanish into smaller-than-expected niches. Natural Language Understanding systems were part of that early wave, with expected applications in automated assistants, innovative computer interfaces, machine translation, and police and military security. The dream has never died, but the applications subsequently seemed somehow subscale against the promise.

Why is this of interest for Next-Generation Networks? There is an astonishing disparity between the two types of traffic NGNs carry: application data traffic vs. audio/video calls. When you point your browser at an application transactional Web site, you have an HTML-based "conversation" with the application where the syntax, semantics, and pragmatics is completely specified. This allows complete automation of each step of the transaction and arbitrary amounts of machine intelligence and formatting can be applied. However, when you use the NGN to talk to another person, to download music, or publish, or observe video content, the network can see the media stream (as a byte stream) but the intelligence it can bring to bear is normally only at the signal layer itself. Typically this is limited to compression for transport efficiency. What a fantastic opportunity for new services and revenues if the NGN could actually understand and produce conversation and video.

As with most advanced technologies, we are not there yet, but neither are we out of the game completely. I will start by looking at automated conversational systems currently in service. I will then look at the most straightforward approach to designing such systems, the so called chatbots, and analyze both how they work and why they are dead-ends in their current form. Next I will look at what has to be done to put together an effective conversational system and some of the reasons why this is hard. Finally, I will outline some of the prospects and their likely impact on the NGN upper layers.

The State of the Art

I recently had a conversation with Andy MacLeod, former CEO of what is now Verizon Europe. I asked him what he thought the hottest issues would be in telecoms over the next few years.

"Going back to my materials science roots, I have to say new kinds of batteries based, for example, on direct methanol fuel cells, which could last maybe two weeks. You go into a shop and buy a top-up, just like with cigarette lighters—that could transform portable devices.

"My next call would be WiFi. It may be cheap and cheerful, but, like Ethernet, it just keeps on improving. With the new mesh architectures allowing scalable municipal networks, both the technology and the economics is looking better and better for new Internet-based access networks. I think WiMAX is really going to have a hard time given WiFi's market momentum.

"For my final candidate, recall the *Star Trek* movie where they go back in time to present-day earth. The Engineering Officer needs to use a current-day computer to design something. He's in the offices of the engineering company and he reaches over to a laptop, picks up the mouse, and says, 'hello computer?' Pretty amusing. I have some experience in the area, back when I used to run one of Nortel's speech processing businesses. It would be fantastic to have a system that did what the *Enterprise*'s engineering guy wanted. I don't know if we're near a breakthrough that would get us there, though."

There is a name for the kind of system Andy wants to see, and it's called a Spoken Dialogue System.

Traci

This morning I called a hotel reservation service. Normally, you are placed in a queue, waiting for a human agent to take the details of your request, but Premier Hotels has an automated agent called *Traci* (T).

[Ringing tone]

T: Hello, I am Traci, Premier Hotel's automated reservations assistant. Do you have a member's profile?

Rule m:

```
       (answer): What is your name?
=> (user): <text> -> $name
=> (answer): Hi, $name, what do you do?
```

The chatbot sends "What is your name?" You type in "Peter" (or "My name is Peter" and the first three words are stripped off) and the word "Peter" gets bound to the variable "$name." The chatbot then answers "Hi, Peter, what do you do?" Convincing to some, perhaps?

The first program like this was Eliza, written by Joseph Weizenbaum in 1966 in the style of a nondirectional psychotherapist. It succeeded in fooling many users with its sympathetic responses, prompting a certain degree of horror and disillusion on Weizenbaum's part about the human condition (1984). However, chatbots are of little use in external-goal-directed activity because they contain little knowledge about anything in the world, and have less ability to do anything with that knowledge. At their best, they hold a mirror to the user. For more information, check the *Personality Forge*, a site dedicated to helping people design and run their own chatbots (http:// www.personalityforge.com). It has contributed a number of prize-winning systems.

Putting chatbots to one side, it is time to return to systems currently in service. How do they work, what do they do, and how can they be made to perform better?

Spoken Dialogue Systems—The State of the Art

The army is in a far-away land, fighting a vicious but diffuse insurgency. The prisons and requisitioned barracks are full to brimming with locals, the results of innumerable sweeps through slums and shanty towns. Most of the captives are probably innocent, but how do you tell? Hardly anyone in the army speaks the local language.

In a high-tech army, most problems are believed susceptible to technology. The army uses an interrogation program called GSTAPO1 (General Speech Translation And Production Operation system mark 1). It has been refined in the field, and here is how it works in practice.

The suspect is brought into the interrogation room and strapped to a steel chair. He or she faces a table, bolted to which is a heavy duty microphone. Speakers and video cameras are visible high on the walls, while the floor is solid concrete striped with cracked and stained guttering. After a scene-setting announcement from the speakers, designed to encourage the suspect to cooperate, a pleasantly insistent synthesized voice poses a number of benign, unthreatening questions.

Me: No.

T: That's OK, I can help you make a reservation or check availability. You can interrupt me at any time, and if you need help, say "help" and if you need to speak to someone, say "agent." So, which premier hotel are you enquiring about?

Me: The Andover hotel.

T: I think you want Dover Central. Is that correct?

Me: No"

[Note that this is the hardest part of the dialogue for the automated speech recognition system to handle. The chain has hundreds of hotels, many of which sound similar. "Andover" must have been recognized as "An" (ignored) + "Dover" and "Hotel" must have been mapped to "Central." What the system hasn't communicated at this point is that it is expecting just the name of the hotel or town with no other extraneous words. It doesn't have a problem with "No" though. Now I get coached.]

T: Please say only the name of the hotel you are enquiring about or say the name of the town or city where you want to stay. You can also say, "help" …

Me (interrupting): Andover.

T: I think you want Newbury Central, is that correct?

[This mistake is slightly mysterious as the sounds are very different. However, Andover is a very new hotel, and Newbury is close to the town of Andover, so it may have been a database look-up error.]

Me: No.

T: Please say only the name of the hotel you are interested in staying in …

Me (interrupting): Andover.

T: What's the arrival date?

[Success!]

Me: Tomorrow.
   [The system has no problem with indexicals—relative references, to "date" here.]

T: For how many nights?

Me: Two.

T: How many adults, aged 16 or over, will be staying?

Me: One.

T: And how many children, 15 or under?

Me: None.

[Notice that this is entirely a *Traci*-centric conversation. I only have to utter single word replies from a highly-restricted vocabulary. As a consequence, my answers are correctly recognized and the "conversation" runs smoothly.]

T: Thank you, so you'd like . . . [summarizes the order] . . . Is that correct?

Me: Yes.

T: OK, please wait while I check availability . . . [Confirms available room and price] . . . Which would you like, make the booking, check the availability of a hotel nearby, or change this enquiry?

Me: Change this enquiry.

T: So, which Premier hotel are you enquiring about, or you can just say the name of the town or city?

Me: Do you take pets?

T: I'm sorry, I didn't understand what you said. Could you please say only the name of the hotel you require . . .

[Notice how brittle the conversation is. Any deviation from the built-in script causes the transaction to crash and burn. At this point, the only recourse is to talk to a human being. However, it is unlikely that the ordinary caller at this time will remember that to do so, they have to say "agent" as they were instructed in the first conversation turn. Most people are likely to experience a rapid increase in frustration and hang-up].

Traci works just fine for routine and predictable transactions, and otherwise fails ungracefully. Once callers are themselves trained on *Traci's* limitations, then I am sure they prefer to deal with "her" than to wait for five or ten minutes in a queue. It's really the audio equivalent of filling in a form on a Web page. And the economics of software instantiations replacing human agents explains why *Traci* was waiting for my call in the first place.

Traci can talk about very little other than making hotel reservations for the Premier company. At the other end of the spectrum, there are systems that purport to be able to talk about anything at all.

The Loebner Prize

Dr. Hugh Loebner set up this competition in 1991 as an application of the Turing test for artificial intelligence. Each year a number of programs compete to appear

most human in their responses before a panel of judges (the judges are typi psychology professors). The interaction is like instant messaging: each judge t a message to the chatbot in a pane on the screen, and back comes a reply.

How convincing is the state-of the art in chatbots? Well, the 2005 wi was a program called Jabberwacky. This program is on the Internet (http://) jabberwacky.com/) and anyone can "converse" with it. Here is my attempt "User").

Jabberwacky: Do you live in Chicago?

User: I live in England.

Jabberwacky: Is england full of programmers?

User: It's normally full of rain.

Jabberwacky: Well what are you up to?

User: I'm writing a book.

Jabberwacky: About e-mail addresses?

User: **You seem obsessed with computer topics!**

Jabberwacky: Bots and maze theory in data structures, are our next up tech.

User: **How did you win the Loebner prize?**

Jabberwacky: No. But then you didn't either!

It looks like Loebner will be waiting a while to award his $25,000 system that judges are incapable of distinguishing from a human being

Why are these systems so poor? They work by superficial manipul rearrangement of the text the user types in, plus a few stock, preloaded The idea is that with enough templates and canned phrases, something ing "normal" human responses can be obtained. Typical rules might b

Rule n:
```
       (user): Do you <verb> <object>?
=> (answer): Yes, I <verb> <object> most days.
```

The user types "Do you like ponies" and the chatbot disturbingly r I like ponies most days." It would have worked better if the activit "eat ice-cream." Presumptively-trivial activities like sorting out verb handled by pre- and post-processing stages. The chatbot mechanisr ages user-specific state information. It can ask for the user's name, regurgitate it later as in this rule.

Q1. "What day is it today?
The suspect often would not know, because they had been incarcerated for days, but the system corrects them and asks them again.

Q1a. "It's Tuesday. Now, what day is it today?"

Q2. "What year is it?"

Q3. "What is the name of your capital city?"

Q4. "What is your favorite sport?"

Q5. …

The intent is to calm the prisoner and get them into a routine of cooperative responses. Some prisoners simply refuse to cooperate, or hurl abuse at the microphone. Fortunately, the chair comes with some persuasive technology that could be used to provide encouragement to respond in such cases. Since there is no time limit to sessions, there are significant opportunities for Pavlovian conditioning.

Once the suspect is cooperative, he Is taken through a protocol scientifically designed to complete an optimized interrogation profile: name, address, occupation, family details, religious and political affiliation, personal history, recent activities, and so on. This is cross-referenced in the database with collateral information, and certain responses are "trigger" items that automatically mark the prisoner's status as more or less interesting. Note that GSTAPO1 is conversing with captives in their own language, but completing the interrogation profile in a language that military intelligence people can understand, thus addressing the critical linguistic barrier.

GSTAPO1 is an effective first-stage filter, and allows the great majority of noninsurgents to be released straight back into the community. However, it is inadequate in two regards. First, there is the problem of type 2 errors, "false negatives." These are the bad people who learn how to "game" the system and pretend innocence: GSTAPO1 mistakenly flags them for release. However, setting the thresholds for a presumption of innocence to a very high level simply means that few can pass it. The result is a mass of "problematically-bad" guys who cannot be released and who overwhelm the small number of skilled interrogators fluent in the local language. Second, the people marked as likely insurgents need a more sophisticated interrogation style than the "tick-in-the-box" approach of the current system. Again, if there were enough human interrogators, fine. But there are not. So the military put in a request for GSTAPO2.

To understand why GSTAPO1 was of limited utility, and how its shortcomings might be fixed, we have to dive inside its internals a little. GSTAPO1 (Figure 12.1) is a traditional, state-of-the-art spoken dialogue system.

The speech recognition module is a standard commercial chipset and statistical package that picks up the sequence of phonemes directed at the microphone by the

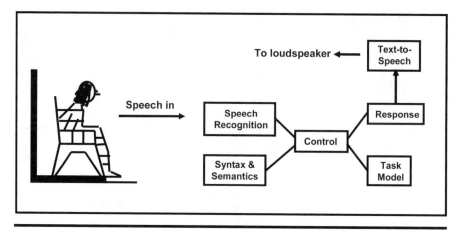

Figure 12.1 The architecture of GSTAPO1.

subject, and matches them against phoneme sequences corresponding to words. Words are not always pronounced in a standard way, unfortunately. Reasons for variation in what is heard at the microphone include:

- Background noise, coughing,
- Regional accents,
- Personal speech idiosyncrasies,
- Stress and fatigue,
- Variability in the time taken to speak the word,
- Mis-starts and hesitations,
- Variant pronunciation of the same word in different contexts,
- Age, gender of speaker.

For these reasons, a straight look-up of "what was heard" in the phoneme-to-word database throws up many possible matches. The next stage of processing, syntax and semantics, narrows these down. The first technique is statistical. By analyzing large numbers of interrogations, it is evident that certain pairs of words have a significant probability of adjacency, whilst other combinations are seldom heard; for examples, see Table 12.1.

Table 12.1 Likely and Unlikely Word Combinations

Likely	Unlikely
threw grenade	few grade
ceiling collapsed	scene claps
no water	nor adder

In practice, we don't try to model the whole of language, just the collection of words that are likely to be relevant to the task at hand, namely primary interrogation. GSTAPO1 uses this kind of statistical model and also something called a contextual grammar. This is best illustrated by an example. Suppose the prisoner is asked "where were you born?" There are a number of ways he might reply (supposing he was born in a place called Barin and was inclined to be truthful):

Q. Where were you born?

A1. Barin.

A2. I was born in Barin.

A3. <expletive> Barin <expletive>

A4. Barin, Barin.

A5. Uh … er … it was Barin in 1983.

Assume, based on many prerecorded interrogations, that this is a good-coverage set of responses. How do we pick out the information we need? The answer is via a grammar network as shown in Figure 12.2 (notice that the grammar will also pick-up related responses not on the above list, such as any answers that end with an expletive).

As the reply from the prisoner is acoustically processed, the speech recognizer is looking, bottom-up, for words that match the sound signal, and that statistically are likely to go together. At the same time, the syntax and semantics component is using the grammar of Figure 12.2 to identify which path through the network

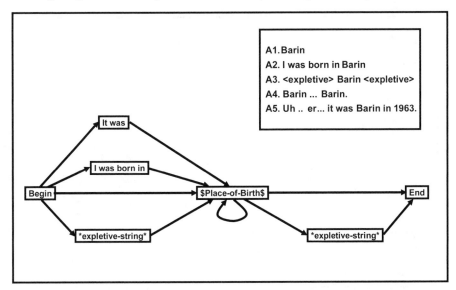

Figure 12.2   Grammar for 'place of birth.'

the prisoner is taking. If there is a way to get the task model variable $Place-of-Birth$ bound to a recognized word (and the system's dictionary will identify words that are place names), then the system will reply:

Q. Confirm you were born in $Place-of-Birth$—answer yes or no.

If the answer is "no," or the attempt to understand the previous question was inconclusive, the system will reask the original question:

Q. Where were you born?

The final module of GSTAPO1 is "control." This structures the overall dialogue. More sophisticated systems keep a track of what they have learned and what they still need to find out, and ask the next question based on some prioritization of what is, as yet, unknown. However, the present system operates a pedestrian preplanned dialogue model—another network, part of which is shown in Figure 12.3.

The dialogue control module starts once the suspect has been made ready. It is a canned speech that simply tells the user what is to come, and how he Is to behave. There then follows the *calming dialogue*, a sequence of benign questions of no intelligence value, but that allow a certain amount of speaker training in the recognition software, and speaker training in the sense of getting the prisoner to a state where she is prepared to answer questions and the system can understand the responses. Then the system gets down to business.

First, there are the standard questions: name, birth-place, address. Each of these questions has its own Q. and A., a contextual grammar as previously shown

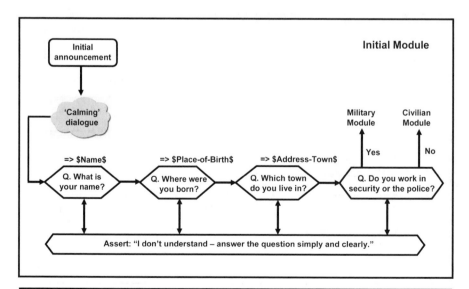

Figure 12.3   Part of the overall dialogue control module.

in Figure 12.2. Successfully recognized responses allow the task model to be updated—the interrogation profile for the prisoner—and lead to further progress within Figure 12.3. If the system cannot understand a response, control passes to the lower box ("I don't understand—answer the question simply and clearly") and the question is reasked.

From the prisoner's point of view, the system is unbearably pedantic, ignoring volunteered information, checking everything with a follow-up yes/no question and taking forever to do the debrief. In intelligence terms, this is a plus, as boredom and repetition helps the debrief process. More commercial systems attempt to accept additional information if it is offered, combine confirmation with further questions and handle a wider range of conversational gambits as in this example.

Q. "Do you want a large or extra-large burger with fries?"

A. "Extra-large and can I have it with extra sauce too?"

Note that "extra-large" refers to the burger, and that the "it" is ambiguous without knowing more about burgers and the menu.

Spoken Dialogue Systems—Raising the Game

The military would like to automate dialogues like the following:

Q1. "Where were you on Thursday evening?"

A1. "I was at home."

Q2. "You were not. You were seen in the old town working on a truck. What were you doing?"

A2. "Did you say Thursday?"

Q3. "We know you were there. <X> has told us everything."

Why can't GSTAPO1 handle this kind of dialogue? Because this conversation isn't preplanned, it's more like a chess game between two players (but with an open-ended set of pieces). To play you have to know a lot about how things work in the world (places, times, travel, trucks, bombs, etc.) and a lot about motivations, why people do things. You also need to make an accurate assessment of what point the other person is trying to make at each stage of the conversation (their move, if you like) so you can find the right conversational countermove with a view to getting an admission. What, for example, is the interrogating party meant to make of this response:

A2. "Did you say Thursday?"

And even the standard problems still exist, waiting to trip the system over. It's easy to say that we won't worry too much about sophisticated syntactic processing of utterances, because people don't speak grammatically anyway. True, but then someone answers like this.

Q. "Who entered the base?"

A. "The men from the militia with the bombs."

Noting as an aside that we got back a noun phrase, not a sentence (ellipsis), what does it mean? Figure 12.4 shows two parses of the phrase, one with the men belonging to a militia that had bombs, the other having the men from the militia themselves having the bombs. We can't tell which is meant, but unless we know there is ambiguity, we can't understand what was just said, and we can't ask the right follow-up question. So it seems we do have to build-in quite a bit of grammar knowledge.

When the military asked for the development of GSTAPO2 to undertake this kind of sophisticated mixed-initiative dialogue, military R&D told them that what they were requesting was impossible. It is beyond the state of the art to:

■ Deploy such a wide competency in language, grammar, and meaning,
■ Build in the encyclopedic level of real-world and situation-specific knowledge required,

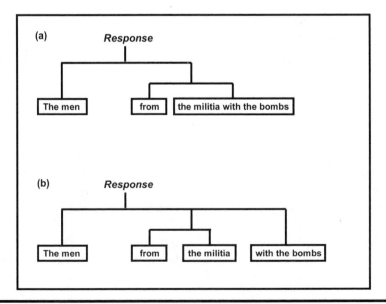

Figure 12. 4 Ambiguous syntax.

- Model and infer the suspect's beliefs, desires and intentions accurately,
- Understand the nature and structure of dialogue itself, sufficient both to understand the real import of what is said and to construct effective replies,
- Exhibit expert-level interrogation skills.

The R&D officer wearily informed his audience that the system they were asking for would not only be able to pass the Turing test with ease, but would also be an effective military interrogator! This combination of tasks would defeat most *people*, let alone present-day Spoken Dialogue Systems.

GSTAPO1 is fictitious (as far as I know), but is based on the architecture for spoken dialogue systems described in McTear (2004) chapter 4. The issues with GSTAPO2 are discussed in more detail in Mitkov (2004), especially chapters 4, 35, and 37 (672). Professor Stephen Pulman drew my attention to http://language.cnri.reston.va.us/TeamTIDES.html where the military are working on technologies that would make such systems possible.

The View from Oxford University

For a clearer view of where we are with language understanding, I discussed the current state of the art in conversational systems with Professor of General Linguistics at Somerville College, Oxford University, Stephen Pulman. Stephen was enthusiastic about recent developments in computational linguistics and this was part of our discussion.

Stephen Pulman: We have an enormous amount of textual information available on the Web, and very powerful syntax resources and processing engines. We can look at how certain nouns or verbs are used in context in Web sites across the Internet, and begin to classify relationships such as "is-a" and "part-of" in a meaning hierarchy using statistical clustering. From a few starter examples, such as "Canada is a country," "Spain is a country," there are systems that can automatically fill-out the relationship, generalizing to other countries.

This may not sound too exciting, but, for example, I expect soon to be able to type into a search engine something like "find me a good price on an Epiphone" ...

Me: What's an Epiphone?

SP: "... they make guitars—and the system understands the concept of "guitar-making company" and can perhaps also suggest something from Fender or Gibson. I expect question-answering to be another area where some new competencies will be on display quite soon.

Me: Like "Ask Jeeves"? Those kinds of systems, as I recall, were pretty hit and miss—really just keyword search.

SP: And they still are, but once a search engine has built a complex hierarchy of linked concepts, then new kinds of more intelligent search suddenly become possible. You could type "When was President Nixon elected?", assuming you wanted to know, and actually get the answer, rather than a list of Web sites that happened to include those keywords, which is what happens today.

Me: What about other kinds of applications, for example, the automation of call center agents?

SP: Well, I don't know the quality of your conversations with service staff over the phone, but it seems to me to be anything but straightforward usually. The conversations always seem plagued by misstatements, misunderstandings, and perhaps some emotion too.

Me: If we can't get it right with people, I suppose there is no hope for automated systems.

SP: Well, the limited telephone bandwidth doesn't help. You'd be surprised the difference it might make to add a video connection. Once we can make a video call to the call center, then the far end can see our face and our lip movements—even our gestures. Research has shown that this extra information can substantially improve accuracy.

Me: Are there any other applications you can see coming along?

SP: You know, I used to be surprised by how few applications there were for genuinely interactive natural language systems. I used to think this was due to the tedium of communication through a keyboard ...

Me: I guess instant messaging and texting might be counter-examples to that?

SP: ... perhaps, but I now think the reasons might be rather deeper. Somehow there needs to be a sense of talking with a real person, a presence. Perhaps we need to link the spoken dialogue systems with household robots to create an "embodied agency," make it real.

Me: You mean an artificial person, or a child?

SP: It's not totally far fetched, there's an EU research project looking at precisely that. It's called CoSy (Cognitive Systems for Cognitive Assistants) and it's looking at integrating many subdisciplines within AI to put together a robot capable of acting and communicating with understanding, perhaps indeed at the level of a small child.

Me: Sounds expensive and difficult.

SP: Maybe so, but I suspect there is a real market for AI-based assistants and conversational partners, particularly with an aging population. It's an interesting question how much better we have to get than the current generation of chatbots so that an embodied conversation agent would be a genuine boon to people needing support. After all, most dialogue is about maintaining social relations rather than answering questions or solving problems. This is an area where we really need to resolve some tough issues to figure out how to do it well enough to be effective in practice!

Conclusions

Putting the scientific questions of theoretical linguistics to one side, the practical engineering of conversational systems has had some successes.

- Call center agent-replacement systems are in service today, although restricted to fixed-dialogue standardized functions such as booking flights and hotel rooms.
- Chatbots with a wide but superficial language skill have achieved some level of dialogue competence, and there are some business models struggling to get launched, for example, in language-learning practice.
- Dictation systems have found a market, and after user-training have achieved astonishing accuracy levels.

Current research is leveraging extensive banks of lexical, syntactic and concept-organization material available over the Internet to induce large-scale concept hierarchies. These will find their use in making search engines more powerful and supporting new kinds of queries based on knowledge and inference.

Some of the bottlenecks to achieving full Turing test competence include the lack of progress in understanding how to capture the meaning of conversation, and the difficulties of understanding exactly what is involved in participating in human dialogues. Is progress here dependent on having embodied systems available that humans can interact with as part of an extended social grouping?

Looking ahead, it seems likely that progress will be more focused on niches where conditions are susceptible to rapid progress, rather than some across-the-board advance to a new paradigm of human-system interaction. But there again, research has a habit of being unpredictable.

References

CoSy project Web site. http://www.cognitivesystems.org/.
McCorduck, P. 2004. *Machines who think*. 2nd ed. Wellesley, MA: AK Peters, Ltd.
McTear, M. F. 2004. *Spoken dialogue technology*. New York: Springer-Verlag.

Mitkov, R. (ed.) 2004. *The Oxford handbook of computational linguistics.* Oxford: Oxford University Press,.

Weizenbaum, J. 1984. *Computer power and human reason: From judgment to calculation.* New York: Penguin.

Recommended Reading

Many aspects of speech understanding and dialogue systems are currently making rapid progress, driven by the existence of more powerful computers, online Internet tools such as corpora, grammars, and knowledge bases, and demand pull from human-computer interface, search engine and general Internet applications communities.

Mitkov's 786-page compendium contains 38 relatively compact chapters aimed at newcomers to the field (although assuming a mathematical/computer science background). It offers a remarkable coverage, with many pointers to further reading and research.

McTear's book much more focused on systems of this type. He is able to get into practical details of how such systems are structured and implemented, providing a strong reality check for those who might believe that science-fiction systems such as *2001*'s HAL are just around the corner. Why is it hard? McTear will tell you.

BUSINESS STRATEGIES IV

Chapter 13

NGN Strategies for Incumbents

Rents and the Value Net

In the textbook case of perfectly competitive markets, it is said that economic profits are bid away to zero. A generation of students were then deeply perplexed, believing that they were being told that there were no profits in competitive markets. Not so fast!

As a company, you can buy capital on the market. The cost of capital is what it costs to service the debt. If you succeed in developing your business so that your profits exceed your costs of capital, then you are making economic profits. For example, suppose you have developed the idea for an Internet TV device that connects to your TV at home and allows your local stations to be viewed by yourself on your laptop, wherever in the world you happen to be, provided you can connect to the Internet. A device like this exists and is called a "slingbox."

You happen to have $1 million to create a business to make and sell these boxes. The product proves to be wildly popular, and you make a profit of $120,000. At least, that's what your accountant tells you. However, if that $1 million had simply been invested in the bank for a year, then it would have made (say) a 5 percent return, yielding $50,000. The economic profit measures the return you get over and above the best alternative you could have chosen, (the opportunity cost you have incurred), and so is only $70,000. Economic profits are always less than accountancy profits.

The textbooks are telling you that you do not make *economic* profits in perfectly competitive markets, they are bid away by new entrants and the increased competition—you will make returns commensurate with putting your money in the bank. If the competitive market is risky, then your expected returns will be equal to bank deposit returns suitably risk-adjusted—multiply the amount expected to be returned by the probability of getting it.

To make economic profits, the enterprise you invest in must have market power. Then, if successful, it can deliver higher real returns than the bank. This is a property of successful patent-holders, monopolies and usually of oligopolies.

That is why it is true that all companies secretly wish to be monopolies, and merely pay PR lip service to competition. It is therefore the objective of business strategy to guide a company to a place where it will have market power, and will therefore make economic profits (Moore 2002).

The other, and related concept, we need is that of rent (Kay 2004, 284). Again, the economists do it differently. For most people, rent is what they pay to their landlord. For the economist, a company (or individual) obtains rent whenever it can charge more for its product than the lowest price at which it would be prepared to do business.

A classic example is a talented sports person or performing artist. As they develop their skills, they become more and more valuable to their team or recording studio. Initially, their organization gets the benefit in increased sales. But the talented person is a scarce factor of production and cannot easily be replicated. They are therefore in a position to extract rents from their employers. So we see sport stars charging economic rent as they auction themselves on the transfer market. Entertainment stars rage about their restrictive contracts, which do not allow them to charge the rents that they now feel entitled to.

Rents are also seen in interbusiness transactions. For example, suppose the Alpha corporation sells broadband lines on the retail market expensively at $50 per month. It can do this because it has monopoly ownership of the access network, and therefore controls a scarce resource that its residential customers are prepared to pay for at that price (those of them who can afford it). Alpha currently charges a very high wholesale price of $45 per month to other ISPs, which makes it uneconomic for them to enter the broadband retail market. The regulator now proposes to set the wholesale price for the product of $20 per month, the estimated long-run incremental cost to provide. If enacted, the intent is that a competitive retail market would then develop around a price converging to $20 plus a small amount extra covering sales, marketing, and support costs.

Incidentally, Alpha is using a tactic here that regulators call the "vertical price squeeze," whereby an operator with significant market power can raise its wholesale price to its competitors while capping its own retail price. The operator internally balances its books by revenue transfer or by being vertically-integrated and not caring, while its competitors' margins are squeezed or eliminated.

The Alpha corporation currently makes wholesale rents of around $25 per line per month, ($45–$20) from any ISP trying to compete with it through its exclusive control of the supply of broadband, rents that it would lose in a regulated market. It therefore lobbies hard with the government against this regulatory proposal. Unsurprisingly, this kind of thing is called "rent-seeking behavior" and public welfare economists tend to frown upon it as it appears to replace investment for genuine innovation with the mere buying of influence to suppress competition.

These concepts have particular force in a value chain, or value network (Kaplinsky and Morris 2002). To deliver the final product to the customer, a number of stages of production are involved, in which different players are linked via market relationships. Some of these players may have significant market power, and will try to use this to extract rents from their partners. In some cases, rent-seeking behavior will be the catalyst of forwards or backwards integration. A classic case is provided by current developments with the Internet as we will see next.

Rent-Seeking Behavior in the Internet

Figure 13.1 shows the value net for content providers such as Google and Yahoo!, Internet platform providers such as AT&T and Verizon, and broadband customers. How does it work?

As a broadband customer, you pay your monthly subscription for Internet access, and all or most of this fee goes to the provider of the access/core IP transport facilities. In North America, this would be AT&T, Verizon, and other carriers with access networks. Internet companies such as Google and Yahoo! need to connect their servers to the Internet, and therefore need to buy high-speed access pipes. Suppose for the sake of argument they intend to buy them from AT&T or Verizon.

Consumer broadband access rates are set either by the market, or more frequently in a process that includes regulation, because companies with access networks tend to have significant market power. Internet access for large companies is a considerably more competitive market, as they buy dedicated fiber links to one or more of the many carriers with high-speed backbones, interconnected to form the public Internet.

However, laid over the market for Internet access is a market for Internet *services*. Google and Yahoo! are very successful businesses. They attract millions of people to their sites, and these visits can be monetized either directly (e.g., Yahoo! sells services such as premium Web hosting and e-mail) or through advertisements. Advertisers will pay to have access to so many visitors, and the contextual nature of the Web allows advertisements to be more targeted, and so more valuable.

Figure 13.1 is really illustrating two value networks: one for Internet access and another for Internet services. They come together because Google and Yahoo! cannot provide their services without Internet access, and this provides incumbent carriers with a gate-keeping opportunity. As owners of a scarce resource (their consumer broadband access networks), they have the opportunity to charge rents. How do they propose to do so? Actually, carriers have a number

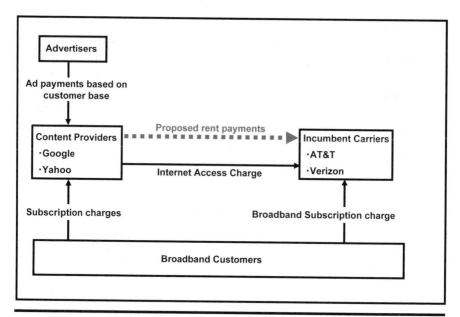

Figure 13.1   Rent-seeking behavior on the Internet.

of different options at their disposal. They could simply try to price discriminate on a per-customer basis. So "rich clients" trying to connect to their networks would simply be forced to pay more. This kind of extortion brought about the common carriage regulations when it was tried by freight companies in the nineteenth century, and it is likely that regulators would be equally harsh today. Or they could simply increase the costs of high-capacity Internet connections disproportionately. However, both these tactics are vulnerable to regulation and competition—there is a more subtle tactic.

At the moment, everyone who connects to the Internet gets a single "best ef- fort" service, but, due to the massive investment in Internet infrastructure back in 1999–2001, this service is actually pretty good. Too good, the carriers contend, as they look for opportunities to segregate traffic into different service classes. The carriers say they will not actively *damage* anyone's traffic as they introduce superior classes of service at various price-points. To admit to anything different of course would be to positively invite regulation.

However, a superior service class *has* to buy something extra, so the most likely scenario is that the carriers will slowly permit utilization levels on the Internet to rise until the resulting congestion separates out an increasingly tardy best-effort experience from superior "gold," "silver," and "bronze" services.

Suppose AT&T tries this. Google says, "Fine. We'll buy our Internet access from one of your competitors who will give us an acceptable best effort service at a competitive rate."

But AT&T has market power. Much of Google's traffic transits AT&T's network, or ends up with AT&T's own broadband customers. AT&T will therefore develop the following policy. On its own network it will institute different classes of service, which will have measurably different performance against standard latency, jitter, and packet-loss metrics. Each service class will have SLAs, and best effort will not be very good. AT&T will then charge its peering partners and connected networks extra to carry any of their incoming IP traffic marked with elevated classes of service.

Suppose Google ended up with a competitive backbone carrier we will call Xcom, which is still on the old agenda of selling best-effort traffic at competitive rates. As soon as Xcom hands best-effort traffic off to AT&T, the quality drops precipitously. AT&T's broadband customers get a very poor Google experience. How happy does that make Google and its advertisers? Xcom had better start marking its traffic and pay AT&T what it wants if Google's traffic is going to have a prayer of end-to-end quality. And how will Xcom then get its money back?

AT&T's market power resides in the fact that its local access monopoly makes it hard for its customers to switch to another supplier of broadband. Clearly, any company merely reselling A&T's broadband service would be at the mercy of AT&T policy here as well, unless regulation was extensive.

The bottom line is that the Internet is finally working as a services platform and real money is being made. The carriers are in a position to charge economic rents for their carriage services, and finally they have a motive to do so—there is money to be made. Their extensive investments in new access networks are real (although they are not doing it to give Google et al. any favors—see below) and these will result in more competition in the wider triple play space, so this is a good argument to use with the regulators. Are the carriers justified, therefore, in violating the principles of common carriage and net neutrality?

Now, there is nothing wrong in principle with offering a portfolio of products at different price points—train and airline operators do this all the time with their first- and second-class tickets. The problem is monopolistic pricing—the exploitation of market power.

The best antidote to monopolistic practices is competition rather than regulation. Recall that the key bottleneck is the broadband access network (there is plenty of fiber backbone out there). If the wholesale price of the access network was held low by regulation, then Xcom, in the example above, could just go round AT&T and offer an end-to-end service at rates below AT&T's proposed new prices—the power to extract rents would be broken. This is unlikely to occur, however, because AT&T, Verizon, and others will not invest in the new access networks unless they get some guarantees that they will be free of such regulation—and they have a point, the new service revenues that would justify these investments are still speculative.

Another alternative is new wireless access networks, WiFi and WiMAX, where costs can be shared, perhaps, with municipal authorities. Google, for example, appears to be buying up dark fiber and is investing in municipal WiFi networks in North America. It is also investing in broadband access through electric power lines. This is a classic example of forward integration by a content provider into the carrier space to avoid being "held-up" by incumbents bent on extracting rents.

Reasons for Access Network Investment

Carriers worldwide are investing heavily in upgrading their access networks to support higher bandwidth services. Some are very expensively taking fiber all the way to the home, others are taking fiber to distribution points very close to groups of homes, and then connecting to short copper loops into individual houses, on which they can run VDSL at rates up to 100 Mbps.

The dominant rationale for this carrier investment is competition from cable and satellite companies who today monopolies revenues streams associated with TV content, and who are increasingly expanding into the triple play portfolio of TV, high-speed Internet access, and voice services. With their new access infrastructures, the carriers hope they can match, and perhaps even exceed the capabilities of their competitors' networks. However, the future market success of carrier triple play services is far from assured, as we will see in chapter 15.

A Twenty-First Century Network

It is not just the access networks that are receiving major investment. Across the world, carriers are contemplating the replacement of their existing voice, leased line, and data networks by an IP/MPLS platform with SIP-based session services layered on top. BT (British Telecom) was the first incumbent carrier to commit to a complete transformation program with its 21st Century Network (21CN). A schematic architecture of 21CN is shown in Figure 13.2.

Like most carriers, BT offers a wide range of consumer and business voice and data products including:

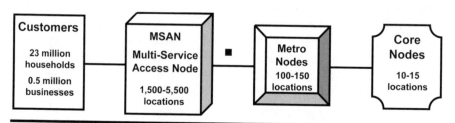

Figure 13.2   Schematic view of BT's 21st Century Network architecture.

- Switched voice (POTS) and premium voice services,
- Transmission services (64 kbps—2.5 Gbps: fiber, with some copper),
- Data services (X.25, Frame Relay, ATM—on copper and fiber),
- IP services (Broadband over copper, fiber, Internet, VPN).

These are today supplied by a number of separate physical networks. The basic approach of 21CN is to retire these separate networks, replacing them with one unified IP/MPLS network (although BT is pushing hard on new Ethernet standards such as PBT, discussed in chapter 2, which could allow Ethernet to replace MPLS).

The product portfolio will also be developed, with so-called new wave services being introduced, but BT sees a need to retain many legacy services for a period. That means that these will somehow have to be adapted at the edge of the new network to be carried across the new core. This adaptation, aggregation and consolidation function is to be carried out by a new kind of device called a Multi-Service Access Node (MSAN)—a functionally-extended version of today's Broadband DSLAM (Digital Subscriber Line Access Multiplexer). The MSAN will support the following functions:

- Broadband IP delivery (ADSL2+, SDSL and forward evolution),
- Data services (ATM, Ethernet),
- Analogue and ISDN voice conversion to/from VoIP,
- Media and Signaling Gateway,
- Low-rate transmission cross-connection,
- SDH multiplexing and cross-connection,
- Wave-division multiplexing.

The MSAN connects to a Metro Node. In fact there is quite a large fan-in from hundreds of MSANs, some daisy-chained, into one Metro Node. If the MSAN is a multi-service layer-2 traffic adaptation and aggregation point, then the Metro Node is where the heavy lifting on service processing occurs. The functions here include:

- Routing (Internet traffic) and GGSN (mobile routing),
- IP VPN Provider Edge VRF functions,
- Firewall, NAT and security,
- BRAS functionality for Broadband connection management,
- Bandwidth management (policy enforcement),
- VoIP media and signaling gateway functions,
- Layer 2 switching (Frame Relay, ATM, Ethernet),
- Cross-connection (PDH, SDH),
- Optical cross-connection.

The Metro Node manages the following service adaptations:

- Analogue/ISDN voice is converted to VoIP (also an MSAN function).,
- PDH/SDH services are encapsulated within MPLS pseudowires.
- Frame Relay and ATM are encapsulated within MPLS pseudowires.
- The Metro Node is also BT's preferred location for interconnect with other carriers.

Long-distance traffic is forwarded to the high-performance core. BT envisages 10–15 core nodes, containing very large routers specialized for brute-force forwarding, the so-called P routers in the RFC 2547 architecture. These core routers will also be positioned at Metro Nodes. The core will also comprise switching at the optical layer, and, for a certain period SDH cross-connection (VC-4 level), while SDH remains in the network.

Above the transport layer just described, BT will implement further functions:

- Session signaling,
- Security,
- Authentication, authorization and accounting (AAA),
- Application services based on Java EE/.NET,
- Bandwidth management and QoS assurance,
- Billing,
- Re-engineered BSS and OSS.

IMS provides standard interfaces and functional components for most of the above.

The carrier concept of the Next-Generation Network is being standardized in bodies such as the ETSI TISPAN group that will flow into a global ITU-T process. The NGN specification will go through a series of releases in which different NGN subsystems will be standardized. These include:

- Network Attachment Subsystem (NASS)—IP address allocation, AAA functions, and location management at layer 3.
- Resource and Admission Control Subsystem (RACS) for call admission control and bandwidth management
- The IMS itself (IP Multimedia Subsystem) enhanced for DSL, WiFi, WiMAX.
- A PSTN/ISDN emulation subsystem, permitting TDM equipment replacement, while keeping legacy terminals in place.

See portal.etsi.org/tispan/ for more details.

Assessing the Program

BT claimed that 21CN will deliver both enormous cost-savings and the most advanced network in the world for new services. They point to the consolidation of equipment, the simplification resulting from the removal of numerous legacy networks, and the flexibility and degree of automation enabled by 21CN.

Critics have pointed out that 21CN is pushing the state-of-the-art in a number of key areas. In some key layers, such as legacy service adaptation to MPLS, only IETF Internet drafts have been available and BT's requirements have been a significant driver. Might it not have been better to wait a couple of years, they ask, for standards to mature?

I was interested in Bob Partridge's view. Bob was a colleague of mine at the Mentor consultancy, and had previously worked for BT as Director, Network Policy, Planning & Performance. Back in 2001, when BT were first considering a fundamental network transformation, they had turned to Mentor and Bob to produce the initial concept and plan. I asked him what he thought of the way BT were going about their NGN transition program.

"The last really big replacement program BT did was the analogue-to-digital switch conversion. This was when they threw out their old Strowger, Crossbar, and Reed Electronic analogue switches and put in the new digital System-X and AXE10 switches. The last Strowger was replaced in June 1995 and the analogue replacement program completed in March 1998.

"Once a factory-like process was going, BT was able to modernize four exchanges per day, achieving a peak of around 3m lines of replacement per annum. However, establishing such a factory process took considerable time as techniques and tools were streamlined and improved to increase the cutover rate.

"A key feature of the analogue-to-digital conversion was that all customers saw massive immediate improvement in their service with touch tone signaling, reduced line noise, shorter post-dialing delay and the availability of supplementary services. All these are now taken for granted but 21CN does not appear to offer any similar direct customer improvements, merely a promise of new, as yet undefined, services.

"The new digital exchanges were complemented by a completely new operations support system which has subsequently been developed over the years with links to the CRM and other systems to provide a high degree of automation of basic functions like number allocation, service initiation and line test.

"The operational and changeover challenges faced by 21CN are therefore very different to those faced in the analogue to digital conversion, with many more internal system interfaces and processes to be accommodated. Additionally, much more complex and larger scale interconnection with other carriers has to be handled and this presents a commercial as well as technical minefield for BT. "

"Are you saying that 21CN is too forced?" I asked.

"Well, BT talk about 21CN as their passport to a world of 'new wave' service revenues and lower operational costs. But a large part of the reason for urgency is the desire to cut operating costs by reducing the number of discrete networks (e.g., ATM, Frame, Digital Switch, Private Circuit), which each carry their own support contract, maintenance, spares, repair, OSS, and inventory management costs.

"Although their existing digital switch infrastructure is considered obsolescent with suppliers not wanting to continue to support it, it remains very reliable. Inevitably it will start to suffer increasing levels of hardware failures, but predicting when this will become a major liability is difficult.

"Our recommendation in the study we did was to adopt a phased approach, concentrating on the most commercially critical areas first and milking the digital exchange asset base for as long as possible in the other areas. However, BT was not enthusiastic about operating parallel legacy voice and new networks for any length of time, so they decided to go for broke with the shortest inter-max period."

"Inter-max?" I had not met the term before.

"Inter-max is the period when BT would be paying to operate both legacy networks and the new network in parallel. They would have the workload of running the business on the old networks, whilst building and transferring customers to the new one. It is the maximum cost period that occurs until all the legacy is closed and the benefits of cost reduction can be taken."

"So do you think the cost savings will materialize as BT argued?"

"I would love to see their current business case! You have to remember that there is little scope for savings on duct, fiber and copper, and for various reasons neither is the scope for economies in buildings and other physical facilities enormous. Many of the digital switches are significantly depreciated and their general maintenance costs are quite low. Additionally, there are all the DSLAMs they have put in as part of the Broadband rollout. These are mostly very new, but they are not conformant to the 21CN MSAN requirements. They don't support VoIP and Media Gateway functions, for example. Are they going to put them all in the skip?"

"Well, I guess not. What about other carriers in the UK, do you think they will all make a similar transition to NGN?"

"Their first major problem is that they are cash-strapped and their network cost reduction opportunities are so much less than BT's because they don't have so much legacy or geographic coverage.

"Second, the new revenue implications of NGN are not particularly attractive. For example, for a major customer, it is unlikely that an all-IP tailored network solution will generate as much revenue as, say, a legacy package comprising discrete frame, private circuit, and ATM products. Potential new incremental revenue also looks to be thin on the ground and the current vicious price competition seems unlikely to disappear.

"Third, many of the alternative carriers have been unable to invest as much as they wanted to keep their existing systems and processes effective and efficient. To gain full benefit from the NGN transition they would need to radically overhaul their associated processes and systems just as BT are doing. Inevitably this will take considerable time and money and as a result, despite the need, the NGN business case is unlikely to be really compelling.

"Finally, consider interconnect. For every non-BT carrier, a major part of their business is interconnection with BT. Most of the traffic they carry will originate or terminate on BT's network. This is due to BT's sheer market penetration. Most competing carriers have tried to connect to BT as low as possible in the switch hierarchy—at local exchanges rather than at the transit layer—thus minimizing BT's backhaul charges. You end up with hundreds of interconnect points, though."

"So will 21CN change things?"

"Well, our recommendation to BT for NGN was interconnection at as few locations as possible to try and reduce the costs of managing hundreds of points of interconnect. They eventually opted for interconnection at the Metro Nodes which are where the large Provider Edge routers are placed, and where they have scale and functionality to provide secure interconnection."

I nodded to myself. The problems of interconnect in NGN are complex, spawning a new industry of session border controllers managing firewall functions, protocol conversion, session proxying, and topology-hiding. Add in the billing, surveillance and performance management functions and this was not a constellation of functionality you would wish to replicate too widely.

"Finally, Bob, what do you think the big issues will be for BT over the coming years of the program?"

"I have no doubt that BT will get 21CN to work. There may be issues of standards maturity, timing, operational problems and payback period, but there seem to be no obvious mega show-stoppers. I also think that they will have their fair share of issues on security, authentication and authorization for the emerging new services. This is very much unknown territory on the scale BT is attempting. It will also be interesting to look at their billing strategy, especially for voice. They are hemorrhaging voice revenues to the service providers like Tesco and Carphone Warehouse and to the mobile companies: I wonder whether the new wave services can possibly compensate and I wonder how they can survive as a major player without a large scale mobile business."

I then asked about NGN scale issues: "People argue that the transition to a next-generation network involves an enormous one-time capital cost, perhaps beyond the abilities of any carrier in the UK apart from BT. But why would this be true? Surely, it's just a matter of extending the IP network—that all carriers have anyway—and simply buying new technologies like IMS, as they come along, as part of normal CAPEX?"

"That might be true for a genuinely new carrier, one without legacy, but most current alt-net carriers are struggling already. The problem is that running a truly national network with a minimum of useful services while retaining enough financial resources to innovate is so expensive that a country of 50 million people can probably support only one such player. Even the United States seems to be emerging as a duopoly, with AT&T squaring up against Verizon. Admittedly they have the new-look cable companies as competitors.

"Logically Europe does have the market size to support genuine competition between viable players, perhaps France Telecom, Deutsche Telekom, and BT, with Telefonica in there somewhere. However, the constraints of 'National Champions,' national security, and civil-service style employee contracts act strongly against consolidation."

The Business Case for a NGN

I had met Mick Reeve, BT's chief architect, at various events over the last few years. We had attended the odd conference together, and participated in negotiations between BT and Cable & Wireless back in 2002. Mick's responsibilities in BT included 21CN, and I was interested in his views on the business case for NGNs. Mick started with his views on services pricing.

"Everyone is facing a tough dilemma on 'new wave' services. Take SMS, the 'short message service.' This is a significant revenue stream for the mobile guys who make, say, 10p (18 cents) for sending only a few hundred bytes. VoIP is, say, a megabyte every two minutes, while sending MPEG-4 video is around 15 Megabytes per minute. If we charged people by bandwidth at the SMS rate, then a two-hour video would cost £600,000 (about $1 million). Not many takers!"

I recalled that one of the stated drivers for IMS was the ability to link sophisticated charging mechanisms to service management, so that charges could relate directly to services, not to bandwidth or bits per second. Mick supported that but pointed out a caveat.

"This works provided we do charge for the service and don't end up providing the bandwidth in a way that it could be re-used for lower bit-rate services. That is essentially why you see the great debates on net-neutrality right now."

What about business models? Did Mick accept that the fate of carriers was, over time, to be relegated to pure bit carriers, a utility business, while the real margins were made by systems integrators in the business space, and content providers in the consumer space?

"I think the value chain is a bit more complicated than that, and also rather service dependent. Take the speaking clock—an example you may think is amusing, or even frivolous. Over the years you would be surprised at how much that has paid for in BT. So with the speaking clock, we are the content provider, but that won't be the norm going forwards. There will be a business model where we do everything for the content provider bar providing the content. We will provide the platform, the ingestion, play-out and management systems, and the billing. And then there is a further business model where the content platform is provided by our upstream customer, and we provide the transport network."

"And the NGN provides you with the capabilities for each and every one of those business models?"

"Exactly! But I would take issue with you that all the value is in content, and everything else is just a utility business. That's certainly the talk at the moment, but the historical evidence is against it. I think you will find that in a steady-state situation, revenues for two-way session services will be three to four times the size of pure content revenues (Odlyzko 2001). Don't write us off yet!"

This, of course, is the key contribution which IMS is meant to make in the next-generation network. People think IMS is mostly about person-to-person services such as video-telephony and push-to-talk, but IMS could equally well be the selection, session-management and charging engine for a video-on-demand service. This could be a powerful package to offer the content-aggregation sector, particularly if there was a way to tie a conditional access system into IMS (chapter 3).

Mick was keen to move the discussion to the access network.

"The future is clearly fiber to the home, or fiber to the curb and then VDSL to the home. This would then give us dedicated two-way bandwidth into each home at speeds in excess of 100 Mbps. None of our competitors could match it. The cable companies have the benefit of coax to the home, which gives them an initial bandwidth advantage, but it's shared, so sooner or later we could beat them.

"In fact—personal opinion—what would really help us are some very popular, very high-bandwidth two-way services people would leave on all day."

"You mean, like the science-fiction idea of 'soft-walls,' or picture walls showing an audio-visual scene piped in from a long way away? Maybe that Hawaiian beach?"

"Perhaps, if it was sufficiently personalized. That would certainly play to our strengths, and would not be easy for our cable and satellite competitors to replicate!"

I wondered if pushing fiber deep into the access network was finally becoming affordable.

"Not at current prices. There's an old network planning rule of thumb that says that the network cost is 20 percent in the core, 30 percent in the OSS and 50 percent in the access. If we just rely on current broadband prices to fund it, it will never happen, or at best take a long time.

"This explains why AT&T and Verizon are pushing so hard to charge for the value-added services enabled by their new network builds. They really have no alternative to trying to 'internalize the positive network externalities.' And if those like Google and Yahoo! think that it's so easy to develop an alternative end-to-end network, then they should reflect on that 20/30/50 ratio. Buying up fiber in the backbone is not so hard: the real expense is duplicating an access network, and the cost factors are much the same whether it's fiber or wireless."

I asked Mick whether he was optimistic or pessimistic about the impact of NGN on BT.

"Completely optimistic. It will certainly lower our costs, but the main thing is that it allows us to engage with many new service providers and service opportunities. We can provide integrated services at many points in the resulting value chains and I am convinced that there's a lot of value there for us. I already mentioned the various business models the NGN opens up, and the importance of new two-way services. If you want to look at us as a pipes business, then fine! The kinds of pipes the NGN will give us promise a very good business going forwards!"

Conclusions

The next-generation network is usually thought of in terms of its technology and service capabilities. However, from the point of view of business strategy, the relevant context is that of the value net.

The current-generation network predominantly supports business and residential communications services, both fixed and mobile. The next-generation network will both support and extend these communication services within the existing value net, but it will also serve as a new IP-based *content* distribution platform.

This new platform will be contended territory between carriers, and content providers and aggregators such as Internet portals, and satellite and cable companies, who have grown used to the idea that distribution platforms are a subordinate component of the value chain that they can control. The carriers will not wish or expect to be commoditized so easily, as Mick Reeve pointed out. Expect both forward and backward integration moves to deal with 'hold-up' problems and drive innovation, and plenty of attempts to extract rents between the various players.

References

Kay, J. 2004. *The truth about markets*. London: Penguin.
Kaplinsky, R., and Morris, M. 2002. A handbook of value chain research, Institute of Development Studies, University of Sussex. http://www.ids.ac.uk/ids/global/pdfs/VchNov01.pdf.
Moore, G. A. 2001. *Living on the fault line* (rev. ed.). London: Collins.
Odlyzko, A. (2001). Content is not king. *First Monday*, (February 2001) Vol. 6, Number 2. http://www.firstmonday.org/issues/issue6_2/odlyzko/

Recommended Reading

John Kay's book *The Truth About Markets* (2004) has a detailed discussion about economic rents in chapter 24. Overall, the book emphasises the roles of "disciplined pluralism" and "incentive compatibility" in the operation of effective market economies—broadly speaking, effective competition, and the alignment of agent interests with desired social outcomes. Many of his points illuminate the issues discussed throughout this book: market structure, the problems of effective competition, bureaucratization, and impediments to change.

Chapter 14

NGN Strategies for Alternative Network Operators

Introduction

This chapter is about facilities-based alternate network operators (alt-nets), and how they should assess the challenge of the next-generation network. Should they invest to remain a complete facilities-based carrier, or should they develop their position in value-added sectors, and perhaps buy-in more commodity network services?

Following on from the discussion in chapter 6, we first take a look at the standard segmentation of the telecoms market, identifying the needs of each segment, and the opportunities and difficulties of doing business there. Second, we review the situation alt-nets currently find themselves in. Does current market positioning make sense, and does it constitute an optimal business strategy for going forward? Third, we look at how needs are evolving across market segments, and what capabilities the next-generation network will bring to the table. Fourth, we look at the value chain in an NGN world and ask where alt-nets could and should play. This discussion centers around the search for premium returns, examining the opportunity costs of prospective investment decisions in an environment dominated by the incumbents, and the many different kinds of players contesting key market areas. Finally, we develop a framework of viable choices that can be used to make decisions.

Market Segmentation and Segment Requirements

The usual segmentation of the telecoms market is as follows:

- Multinationals,
- Large enterprises,
- Medium enterprises,
- Small enterprises,
- Consumers,
- Wholesale.

There is often a more detailed segmentation of the business segments, large and small, into vertical markets such as finance, technology, retail, and so forth. In this chapter, the focus will be on business segments considered horizontally. The next chapter looks at the consumer market. There is also the inter-carrier wholesale market, which is often a significant part of the business by revenues, mostly for commodity services. In what follows, it will mostly be discussed from a buyer rather than a seller perspective.

In the past a company's IT needs and voice/data communications needs were considered largely separately, but with increasing technology and service convergence this is less and less true. We now talk about ICT (Information and Communications Technology) as the basis of the customer proposition.

Business customer requirements for a *holistic* ICT solution create stresses for the existing value chain. Carriers, while expert in networking, typically know little to nothing about IT; IT systems houses and systems integrators typically do not own networks. This creates a contested space where the two sets of providers overlap in providing ICT solutions. Carriers are conscious of coming off worse, as network services can more readily end-up as barely-differentiated near-commodities. Premium margins then appear to accrue to the IT-based systems integrators.

A large part of the recent history of carrier business models is an attempt to escape this commodity trap. The NGN is widely seen as a multi-layered platform that can potentially bring more premium value back into the carrier space. Whether carriers are up to the task of exploiting it in this way remains to be seen.

Multinationals

Multinationals tend to work with bigger carrier players, and because of their buying power and substantial in-house integration resources, are often commodity purchasers of basic connectivity services at deeply discounted rates.

More recently, many multinationals have teamed up with global systems houses such as IBM Global Services and EDS. The buying power exerted by these systems integrators is, if anything, even more powerful than that due to the multinationals themselves. Carrier margins are therefore further squeezed.

While there are reasons to go after this segment of the market—mostly to do with the sheer size of the business even at low percentage margin—it tends not to be a major priority for alt-nets, particularly if the alt-net is only national in scope.

Large Enterprises

Large enterprises have been a "sweet spot" target for most carriers. They are few enough to justify dedicated account teams, complex enough to require customized and bespoke solutions, and rich enough to pay for them. Consequentially, the margins can be good if the proposition is right.

The large enterprise market segment exhibits the already-mentioned contested space between carriers and systems integrators. Carriers like to talk about their partnerships with preferred systems integrator collaborators. The truth is, this largely reflects the weakness and difficulty carriers experience in developing competences in IT systems and professional services themselves. However, these are just where the margins are, so carriers never give up hope of forward-integrating into these areas. The opportunities and difficulties in so doing will be discussed in more detail later in this chapter.

SMEs

Small and medium enterprises are often lumped together as SMEs. A "small enterprise" is usually one without dedicated IT/technical staff: you deal with the owner/manager who is often completely nontechnical. A "medium enterprise" is large enough to have IT/networking staff, albeit only a few of them: you talk to a technical person who understands your portfolio.

SMEs they have typically proved hard for carriers to address. There are so many of them that dedicated account teams are impossible. They cannot afford customized products wrapped around with significant systems integration. But it has been difficult for carriers to design and deliver standard product building blocks that can be cheaply configured in this sector. The problem is often handed across to systems integrator (SI) partners and value-added resellers (VARs) who specialize in cost-effectively addressing the needs of these customers. The VARs succeed because they are customer-solution centric, and can construct their solutions from a wide range of suppliers.

There was a time when it was believed that Internet self-service was the answer to cost-effectively addressing the SME sector, but a scalable strategy has proved elusive. ASPs like salesforce.com have had niche success (also with larger customers) while some ISPs have created highly automated hosted services for SMEs who are sufficiently technical. Attempts to broaden the offer continue, with more advanced self-service portals. ICT services are not books, but the example of Amazon.com is never far away. It has to be said that the Internet self-service portal has had considerably more success as a way to smooth interaction with partner VARs.

Consumers

The consumer sector is characterized by the centrality of access. In pretty much every developed country a regional or national incumbent monopolizes the ownership of copper loops to the household. Usually, the only competition in providing a two-way service mechanism is a cable company. Satellite, of course, provides an excellent one-way service, as Sky and DirecTV have proved, but this is not economic for two-way services.

Wireless (WiMAX, WiFi) is also much discussed, but technological immaturity, spectrum scarcity and deployment costs inhibit this alternative to date. It may turn out to be the case that the WiMAX "sweet spot" is actually as a kind of big brother to WiFi in campus applications for businesses in fact. WiMAX is distinguished by having better QoS, a better hand-off architecture (802.16e) for roaming and a longer range than current WiFi.

Finally, 2G and 3G mobile technologies have provided a way to reach consumers, but the barriers to entry for further cellular operators in most countries are absolute, with no further licenses being allocated. MVNO opportunities are still there, limited by the capabilities and cost-economics of the 2G/3G mobile networks. In particular, cellular technologies have not proved a cost-effective way to deliver broadband services to homes to date.

For a long time, the consumer sector was written off by alt-nets. Incumbents held onto control of the local loop, and despite unbundling attempts by the regulator, wholesale products were unattractively priced and hedged with bureaucracy, increasing transaction costs. Alt-Nets abandoned the consumer markets and searched for richer rewards with business customers.

However, in some markets, notably the UK, the regulator has pushed hard for a more competitive playing field. Customers can now make a one time decision as to who will carry their calls, and the incumbent, BT, is then obliged to route the call via that operator. This is called "Carrier Pre-Select" (CPS) and obviates the need for prefix-dialing or special boxes inserted into the line between the customer's handset and the phone socket. Wholesale costs have been addressed through lower line-unbundling charges, and to enforce equality of access, BT's access organization has been reorganized into a separate division called Openreach.

As a result there has been real competition to BT both in traditional POTS via CPS and in broadband provision. In the later case there are opportunities both in reselling BT's wholesale broadband product, and through line unbundling, with the alt-net putting its own DSLAM into a BT exchange, or at a nearby site. The

ity of IMS to be extended to a variety of non-cellular access methods including DSL broadband, WiFi, WiMAX, and cable. There has, however, been little sign of IMS-powered innovation in the kinds of services business customers might be prepared to pay for. The sorry conclusion is that IMS will power a number of future business services, but the elusive killer-application is still just that.

The Business Value-Chain Ecosystem Facing the Alt-Nets

The key to understanding the emerging telecoms value chain, and the options within it open to the alt-nets, lies with the ever-advancing development of ICT service possibilities and the consequential evolution of customer requirements. Companies know they have to create a secure multi-service networked applications infrastructure, and they need help with a list of concerns that can stretch like this:

- Converged intra- and inter-site QoS-enabled communications network,
- Voice and multimedia capability (VoIP, wireless, mobile + integration),
- Evolution of IT platforms to Web architecture on Intranets,
- Contact centers, Internet E-commerce, Extranet, and Storage solutions,
- Application Infrastructure Hosting (Java EE/.NET),
- Use of ASP-provided application services, or managed outsource,
- Security—all aspects,
- Performance and cost monitoring linked to SLA management.

The challenge for alt-nets (as for incumbents) is to put together a collection of hosted products that can address each of the business needs on the above list. In principle, a carrier can do it all, but the case has to be made as to why a carrier-solution is better than an enterprise doing it in-house. One argument is flexibility.

An *a la carte* list like the above can be read as a static set of requirements. But businesses face an increasingly changeable environment. Just putting in place a solution and letting it run ballistically over the next five years won't work. What is needed instead is the ability to set up a solution on day one, and then modify its attributes with low cost and effort on a day-to-day basis.

This need for flexibility creates a new emphasis on the customer *experience* of the service, and for a sophisticated set of instruments and controls for driving it. Carriers cannot achieve this level of responsiveness without putting in place a new generation of Business and Operational Support Systems (BSS/OSS). As companies like Vanco have shown, this can be a key competitive differentiator. It is all very hard for a first-generation alt-net choked by legacy systems and processes.

Companies have a choice of who to do business with. They can choose to stay in-house or deal with systems integrators, who can provide integration skills through their professional services arms, and who can run the company's application and network services via a managed outsource. Companies can also do business directly with carriers who own networks and wish to offer managed services. And behind both these managed offers is the world of vanilla network products—commodity already or rapidly commoditizing. The result is a value chain as shown in Figure 14.1.

In this value-chain, we see a process of commoditization growing from the bottom, as network and application platform components become more standardized and plug 'n' play enabled. We see a growth of the premium managed service market at the top, as more powerful and extensive components, many due to the NGN, drive more sophisticated services.

Incumbent carriers have a reflex to preserve their historically successful vertically-integrated model. BT is a case in point, with the network assets concentrated in the BT wholesale division, managed services for consumers and business in

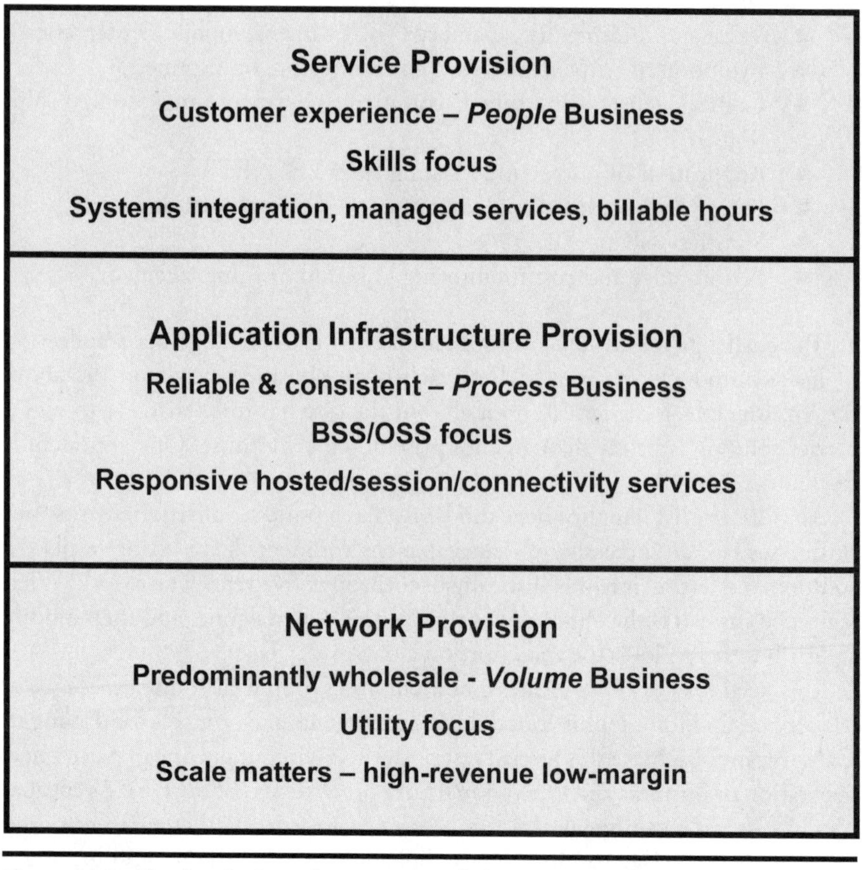

Service Provision

Customer experience – *People* Business

Skills focus

Systems integration, managed services, billable hours

Application Infrastructure Provision

Reliable & consistent – *Process* Business

BSS/OSS focus

Responsive hosted/session/connectivity services

Network Provision

Predominantly wholesale - *Volume* Business

Utility focus

Scale matters – high-revenue low-margin

Figure 14.1   The developing telecoms value chain.

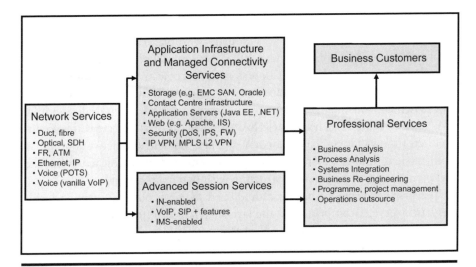

Figure 14.2   Business Services in the ICT Value Chain

BT retail, and BT global services providing business services, integration, and ICT solutions. However, as we look more closely at the services provided across the value chain (Figure 14.2), the disadvantages of vertical integration become more apparent.

The network services utility business on the left is a commodity, scale business, and arguably not best-provided by smaller alt-nets in the presence of a viable wholesale market.

There is still scope for differentiation in Application Infrastructure Provision, Managed Connectivity, and Advanced Session Services. These services are still in technical evolution, and possess many attributes where carriers can differentiate on the basis of technical competence, service quality, ease of use, flexibility and manageability, in addition to price. However, there is a general trend towards lowest-common denominator commoditization over time. Connectivity products such as BGP/MPLS IP VPNs are well down that road.

The Professional Services business also offers scope for differentiation. The ability to engage with a customer and put together an effective ICT solution integrating both insourced and outsourced components requires competence in many skills. The difficulties in achieving a reputation for consistent high-performance lead to premium returns for those who can achieve mastery. Since different verticals often have distinctive requirements, there are many opportunities for niche positioning.

Carriers have historically targeted the ICT services area as the holy grail of their future business. However, the skills needed to be a leader in what at heart is a people business are so discordant with the reutilized process skills needed to be successful in delivering managed platform services that success has generally eluded them.

The situation facing carriers is made worse by the presence of a large, active, and skilled professional services industry with players ranging from global corporations—IBM Global Services, EDS, Accenture, and the like, through to second tier pan-national systems houses such as LogicaCMG in Europe down to national or regional Value-Added Resellers (VARs). These organizations know how to manage consultants, developers and integrators and are as familiar with IT systems from vendors such as Oracle, SAP, and Microsoft as they are with telecom products from vendors such as Cisco, Avaya, and Siemens and network products from carriers. And none of them have to operate and maintain networks. The end-game may well be a market structure with two main camps.

The first camp would be a commoditized *platform* business offering facilities-based services comprising all the layers envisaged in the NGN architecture (including IMS and Java EE/.NET platforms). This would be the domain of larger facilities-based carriers, chapter 6's generalists.

The second camp would be the *managed application services* business offering a mix of insourced and outsourced application services tailored to the sector and business-specific needs of the client. All (commoditized) platform services would be bought in and packaged to form the solution offered to the customer. This would be the domain of the systems integrators, VARs. and ASPs.

However, we are many years away from such a clean "dis-aggregation" model. We should include amongst the factors impeding such a transition both the historical inertia of large vertically-integrated incumbents, and the immaturity of standards and products—especially at the higher NGN layers of IMS and Web services. These complex, not-yet-commodity technical components offer opportunities for platform-based carriers to excel in bringing them to market in a customer-friendly form, and will resist commoditization for many years to come

As a consequence, there will be significant opportunities for carriers in general, and alt-nets in particular, to bid for the contested area of value-added services in the next period. This creates a significant challenge for each alt-net as to where to focus its investment priorities—what should they buy, what should they make, and what should they sell (and to whom).

A Framework for Market Focus and Some Optimal Scenarios

In telecoms, incumbents are frequently former monopolies who, despite deregulation, still maintain a dominant hold on the market, with market shares of 60 percent or higher overall. This leaves the remaining small slice of market pie to be fought over by all the alt-nets.

In the earliest stages of deregulation, back in the 1990s, telecoms was often a duopoly, or a restricted-entry market. This encouraged the first-generation alt-net to be a full-portfolio generalist copy of the incumbent. The sub-scale inefficiencies of the alt-nets in this mode were somewhat compensated by regulation: the alt-nets underpriced the premium charges of the incumbent by 10–15 percent and were still able to make a profit despite their inefficient cost base. The regulator prevented the incumbent from taking its minnow-sized competitors out.

The Internet boom encouraged a lot more national infrastructure build, bolstered by the belief that IP networks would soak-up exponentially-increasing E-business traffic and revenues for the indefinite future. Investment was also encouraged by the belief that the incumbents and first-generation alt-nets, hampered by their legacy networks and systems and their "bell-head" mindset, just wouldn't "get it" and would prove to be weak competitors. We all know where that ended up.

The market structure theories of Sheth and Sisodia (2002), discussed and summarized in chapter 6, predict that given a market dominance of around 60 percent by an incumbent, there may be room for only one further generalist, and after that the remaining pie is too small. The only market strategy that works for everyone else is to be a specialist in a sector defined by geography, customer type, or product type. In such a niche, the generalist can be out-performed, and, with care, a niche monopoly can be established. To hold onto this market beachhead, Sheth and Sisodia recommend the niche player to shun fixed costs, to stay flexible and use whatever tactics are available to lock-out competitors. This is good advice for most business-oriented alt-nets going forward.

A convenient framework to analyze these niche opportunities is Michael Porter's five forces model (1998). As a reminder, Porter analyses the situation of a company in its environment in terms of the following five forces.

1. Competition between companies within the sector (on price, quality or service).
2. New entrants who could take market share or depress prices.
3. Substitute products, and opportunities via complementary products.
4. Supplier power utilized by vendors to extract value.
5. Buyer power, e.g., demand for deep discounts by customers.

Figure 14.3 illustrates a (different) five-step model demonstrating how a typical alt-net would use Porter's framework to help plan its business future as a premium-return niche-specialist, based on Besanko, Dranove, and Shanley (2000).

Step 1. Determine Possible Niches

Given the inventory of assets, products, and customer relationships held by the alt-net at the moment, what geographies, customer-segments, and product-areas could the company plausibly play into? For the sake of this example, we will assume:

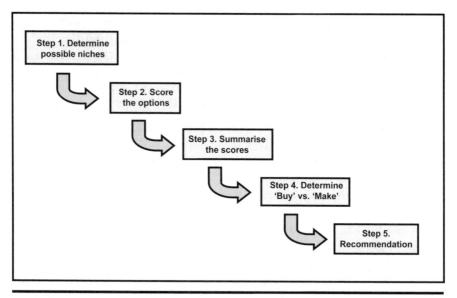

Figure 14.3 Find your niche.

- A national geographical play.
- Possible customer segments: large, medium and small businesses plus a wholesale offer to other carriers, and a retail service to SIs and VARs.
- Possible product-areas given shown in Table 14.1.

In practice, the granularity of analysis need not be as fine as Table 14.1. Products can be clustered in groups that are offered together. The activity of clustering is

Table 14.1 Product Areas To Be Analyzed Using Porter's Framework

Possible Product Areas

1. QoS IP network
2. VPN (various types—L3/L2)
3. IP Centrex (hosted VoIP package)
4. Fixed-mobile convergence package
5. Hosted Contact Centre
6. Hosted Web/application servers
7. Hosted storage
8. Hosted eCommerce package
9. Hosted (vertical) applications
10. Security products (MFW, IPS, AV, audit)
11. Performance, cost monitoring packages
12. Advanced (self-service) management systems

equivalent to defining the separable markets that constitute the possible available niches. For example, providing high-quality VPNs with sophisticated management and reporting functions has constituted a horizontal niche for some companies at some times. Contact centers, which are complex to set-up and integrate, can also provide a niche for specialists.

Step 2. Score the Options

For each customer segment (and geographical area, if there is more than one geographical option) assess each product area above under Porter's five headings—both examining today's situation, and the future (three-year) trend.

An excellent template for this is described in the appendix to chapter 11 of Besanko, Dranove, and Shanley (2000), which demonstrates just how many issues can usefully be considered under each of Porter's five headings. Table 14.2 shows a suitable format, which is reused in Step 3. In Step 2, detailed answers should be written in the second column to questions under each of Porter's headings in column one (which are summarized into *low*, *medium* or *high* risk in Step 3, as we shall see). What are the questions we should be asking under each of Porter's heading?

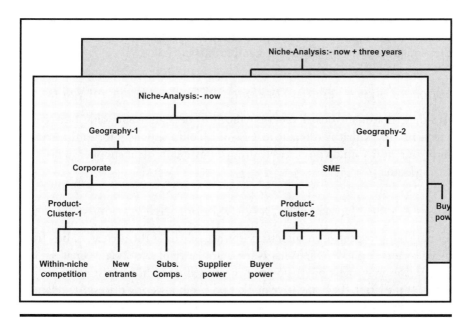

Figure 14.4   Niche analysis structure.

Competition between Companies within the Niche Sector (on Price, Quality, or Service)

Who else is playing? How stable is pricing in this niche and how are prices likely to be set? Are there few or many players? What are the cost bases of the different players? What are customer switching costs? What is the timing and transparency of the sales cycle (which could facilitate or impede price competition). Are you competing on price, quality, or service, and are there opportunities for horizontal differentiation (varying optional discretionary attributes) and vertical differentiation (varying core quality attributes)?

New Entrants Who Could Take Market Share or Depress Prices

Is there a learning curve for this product, and where is the alt-net on this curve vs. possible new entrants? Are there regulatory issues that could admit or impede new players? Would existing players punish new entrants, for example by waging a price war? How important is brand and/or scale for success in the niche? Is the niche broadening in such a way that it could fuse with other product areas, bringing in new entrants from that space?

Substitute Products and Opportunities of Complements

Are there substitutes or complements at all for this product area? What is the price sensitivity in the niche, i.e., if prices change, do customers rapidly seek substitutes or are they relatively insensitive to price? Are there prospective exogenous changes that could stimulate or damage demand in this niche?

Supplier Power—Vendors Positioned to Extract Significant Value

Are there dominant brand vendors who provide significant value but charge premium prices that is difficult to pass on? Could suppliers forward-integrate into your area? Could you be easily dropped by a dominant supplier in favor of a competitor?

Buyer Power—Demand for Deep Discounts by Customers

Could your prospective customers relatively easily do it themselves, or buy from allied sectors (e.g., VARS)? What is the granularity of sales (many small vs. few large)? This may affect buyer power. Are you selling to many customers, or just a few? If the latter, they may have market power in lowering prices. Are relationship-specific assets involved that can be used for "hold-up" by your customer to leverage your prices down? Are your customers likely to be price-sensitive?

As a strategic marketing activity, the alt-net team should first define its grid of relevant attributes as described under each of the five force headings above, put together a list of possible niches, and then score each of them using a grid along the lines of Table 14.2. It would be a good idea to have a column for the immediate situation, and another for the "three-year out" situation.

The value of this exercise is predominantly the process of asking and answering the questions, to get a feel for the opportunity to colonize this geography-segment-product niche at this time and defend it, in order to make premium returns. After all, the next stage is simply going to summarize the team's niche assessment into one word.

Step 3. Summarize the Scores

Once the detailed grids have been completed, they should be summarized for each niche according to Table 14.2 (completed with an assessment for BGP/MPLS VPN). For each prospective niche, for each of the five forces, we estimate the consolidated risks to our ability to make premium returns.

You can disagree with the example assessment, but based on Table 14.2, the opportunities to make good returns in this already-overcrowded market do not look promising. If it is worth entering, it may be as a "table-stakes" enabler for other products where the opportunities are better. Or there may have been a way to tailor the product for a vertical market in such a way as to improve the scorings.

I emphasize that this exercise must be evidence-based. It is important to *itemize* competitors, possible new-entrants, suppliers, customers, and threats from other substitute products—not just guess from gut feelings. Once this exercise has been completed across the relevant niches, it should be possible to identify

Table 14.2   Five Force Niche Summary Assessment

Niche Description

Time = 2007

Geography =UK

Segment = Corporate

Product = BGP/MPLS VPN

Five Force Category	*Risk to Premium Returns*
1. Possibility of new entrants	High
3. Substitutes and Complements	Medium
4. Market power from suppliers	High
5. Market power from buyers	Low

the areas where opportunities exist, and this constitutes the basis of a "go-to-market" strategy.

Step 4. Determine "Buy" Vs. "Make"

The decision to buy components of a product rather than provide them in-house implies that the market-supplied component is at least as cheap, as high-quality and as reliably available as that which could be provided internally. These qualities rely upon the existence of a stable reasonably competitive market.

Some products in the telecoms market are effectively commodities: examples include leased lines at rates from 2 Mbps up to high-speed SDH links at STM-4 (622 Mbps) and beyond. In the UK copper loops have been unbundled, finally at competitive rates, through intense efforts on the part of the regulator.

The monopolistic power of the incumbent implies, however, that when regulation is light, key services will be found to be simply unavailable on the wholesale market, or overpriced, or subject to major transaction costs due to gratuitous bureaucratic obstacles.

For example, in theory it would be possible to lease a national fiber infrastructure from a third party. However, if the alt-net already owns one, its sale value is unlikely to be very large in today's over-supplied market, and continuing to operate it is most likely cheaper than leasing comparable facilities from anyone else. Naturally a more detailed analysis would have to be carried out. Access circuits, where a price-competitive market obtains, may be a different matter, as the alt-net is unlikely to have extensively built-out such a network itself to all points where it needs access.

When it comes to the elements of the next-generation network itself—layers such as the scalable QoS IP network, the IMS layer, advanced application platform hosting, the make vs. buy decision becomes problematic again. It would be best to buy the standard component services on the wholesale market from the incumbent, benefiting from its economies of scale. However, the regulator is likely to adopt a very light-touch with the incumbent over the next few years, as the latter will have made the case that such a risky investment needs some guarantees of return. The wholesale market for services based on these NGN platform layers is therefore likely to be difficult for alt-nets in the near to middle future.

The product mix underpinning the niches of choice from the preceding stage of analysis undoubtedly requires NGN platform elements as input factors. VPNs with QoS require routers and network links that can deliver the required services; VoIP with features demand soft-switches/IMS platforms that deliver the said features; performance monitoring, reporting, and service management requires advanced BSS/OSS systems and the APIs to be available into the managed network elements and servers.

As a consequence, it may be worth the alt-net investing in layers of the NGN, which are required input factors to its preferred services, even if it does not do so at scale. The inefficiencies may be outweighed by premium returns over the whole product, and there may well be additional opportunities for technical differentiation or learning-curve advantages in some of the newer technologies of the NGN. Again, detailed modeling is required.

An entry into professional services is not excluded. There appear to be few cases where an alt-net carrier has organically grown an SI division, but since systems houses exist at all scales, and are frequently quite specialized in terms of product and vendor competencies, acquiring some should not be too difficult. The source of competitive advantage will then come from the ability to link systems analysis, delivery and integration skills tightly with platform technologies and capabilities already owned by the alt-net. This is still sufficiently hard that making it work well can command a premium, even if this will be less true a few years out. Alt-nets should be aware of the dangers of killing SIs after acquisition by constraining their freedom to put high-quality solutions together, by subjecting them to routinist, process-centric management regimes, or by treating them as a loss-leader, thereby underpricing their services.

Step 5. Making a Recommendation

Finally, it should be possible to pull it all together, and make a detailed recommendation covering:

- Geography, customer-segment, and product niches to priorities,
- Product portfolio to be developed,
- Resources to be provided in-house vs. bought on the market,
- Required acquisitions and disposals,
- A costed business case and roadmap.

The time span of such a strategy is probably three years—certainly not more than five.

Carriers and the SME Market

I have long been puzzled by the ambivalent attitude alt-nets have to SMEs. The sector is usually considered to be underperforming, with large enterprises being seen as far more profitable. CEOs complain there is a very long tail of underperforming customers, but it seems difficult to identify and dispose of them. On the other hand, there are many VARs and vendors serving the SME sector and they don't appear to be doing too badly, so what do *they* know which carriers don't?

I was fortunate to be able to talk to a senior executive with a long career in selling to this sector. I asked whether the SME sector was always a poor performer.

"People sometimes think that, but it's not necessarily so. I have seen significantly better returns from SMEs even than from larger enterprises, where costs for bespoke solutions often pull profitability down."

"So how do you succeed in this sector?"

"Small and particularly medium-sized companies today absolutely rely upon networked IT services. They prefer to have a relationship with a company about their own size which they find more affordable and which gives them a far more personal service. This is why they rather prefer to deal with Value-Added Resellers (VARs) rather than directly with carriers most of the time. They're also very nervous about carrier lock-in."

"So do you put together highly-tailored comprehensive products for the VARs to resell on your behalf? Is that the way you make money?"

"Not really, we find that the VARs are often concerned to use the integration skills which they take to be the core of their own businesses, and don't necessarily want us to do all the work in advance. And in any case, SMEs are extremely keen on choice and don't take kindly to having just one overbundled solution put in front of them. We actually find it better to offer a portfolio of focused products to the VARs for them to sell on."

"So where would you say were the sources of your competitive advantage?"

"I would say in two areas. Firstly we make huge efforts to be easy to do business with. For example, we try to facilitate "touch-free" self-service via our VAR-partner Internet portal. The second source of advantage is that we are a new player. We already have a next-generation network, and most importantly a next-generation set of business and operations support systems. Without legacy, we have the right kind of service management flexibility right now, and our cost base is low."

"So the key then is focus?"

"Absolutely. Our focus is flexible managed services aimed squarely at the medium business sector, and that's what we're good at. We may tactically try many things, but strategically, that's where we are."

Summary

Alt-nets are a mixed bunch. Some of the more established alternative network operators are smaller copies of an incumbent, with similar mixes of legacy networks, systems and products. Others, particularly those built in the Internet boom, claim to already be next-generation networks. It is emphatically not the case that one business strategy fits all.

There is usually room for one or perhaps two alt-nets to aspire to be generalists, junior competitors to the incumbent across most market segments and products. Unfortunately, the most likely candidates for this role are the older legacy alt-nets and they are the ones finding the next-generation network transition the most difficult. However, with sufficient capital resources, for example via acquisition by other major players, some will succeed.

Other alt-nets, particularly the newer IP-based players who have not found a deep-pocketed partner, should understand that their future lies in high-margin niches. They will have opportunities to address advanced ICT problems in both the large corporate and SME sectors, but will have to do so in partnership with systems integrators and VARs. There is an overriding requirement to identify a stable differentiating market focus, and then structure investment to shape the alt-net for success in that niche and lock-out competitors. I outlined a five-step process whereby suitable niches can be identified and the necessary "make or buy" decisions made.

There is a slow process of bottom-up commoditization of all network services. BT likes to describe the process top-down as the "sedimentation" of once-premium services to merely commodity status. But with the next-generation network being very new, and new services still in concept stage, this transition to commodity status, however it is called, will take many years. In the meantime, there are opportunities for both alternative operator generalists and specialists, providing the focal points are well-chosen. Key concepts for success are to priorities flexible business and operational support systems (BSS/OSS), without these the alt-net is too unresponsive and immobile to succeed, and to invest in technologies and processes that make it easy for customers and partners to do business with the alt-net. These have more choices than ever, and will not tolerate poor service.

References

Besanko D., Dranove D., and Shanley M. (2000). *Economics of strategy* (2nd ed.). New York: Wiley.

Halabi S. 2003. *Metro ethernet*. Indianapolis, IN: Cisco Press.

Porter M. E. 1998. *On competition*. Cambridge, MA: Harvard Business Review.

Sheth, J., and Sisodia, R. 2002. *The rule of three*. New York: Free Press.

Chapter 15

NGN Strategies for Capturing the Consumer Market

Introduction

Consumer services can broadly be divided into two major areas: *content* services, where the customer is offered professionally produced material for entertainment, information or education; and *communication* services, where the customer is offered the ability to create a two- or multi-way information-bearing channel with other communicating entities (one or more persons, computer systems, or some combination). This distinction is not at all absolute: TV programs come with interactive features such as viewer voting, while Internet surfing is all about accessing content. However, in business terms, the value chains, or value networks, are today largely distinct. And this chapter, as previous ones, is anchored by the concept of value networks.

The Internet and the NGN supply a kind of "external shock" to the existing value chains, both for content and communications services. We will start by considering the former. In this chapter content means digital content such as digital TV and radio, film, and music. Where I mention TV, both the broadcast of linear channels and video-on-demand (VoD), I will usually mean the possibility of radio channels as well. The relevant technologies were discussed in some detail in chapter 3; here we look at the business structure of the market and the strategies of the various players. After discussing content services, we will take a look at the business strategies for players in the *communications* services sector of the consumer market.

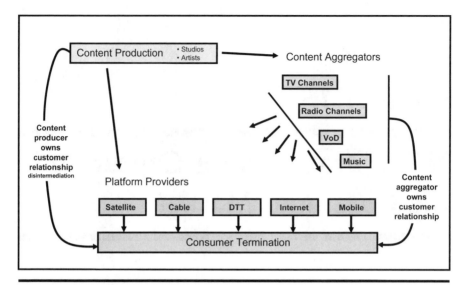

Figure 15.1   Value network for content services.

Business Strategies in the Content Services Sector

The value net for content services is shown in Figure 15.1. It comprises three main types of player: content producers, content aggregators, and platform providers. We will take a look at each in turn.

Content Producers

Content Producers create programs, films, books, music tracks, and performances with the intent that consumers should find their work valuable, and should pay to experience it. The problems they face are to acquire an audience at all, to deliver their product to that audience, and to secure payment. There are so many potential providers of content in a busy and inattentive world, that all of these things can be very difficult. Content producers usually need help from downstream in the value net. However, things are dynamic. If a content producer can acquire a reputation, a devoted (read "locked-in") audience, and a low-cost distribution and payment mechanism, then the said performer might be able to appropriate most of the value from his or her work. Internet bands, at least in the early stages of their career, are a contemporary example. This worries other parties to the value chain who fear "disintermediation."

Content Aggregators

Samuel Johnson famously said, "No man but a blockhead ever wrote except for money." However, the author frequently finds that he or she needs help to monetize their excellent literary content. The purpose of content aggregation is to better secure revenue streams, either from consumers through subscriptions or pay-per-item, or from advertisers drawn by the aggregated audience, or frequently from both. In order to achieve this, the content aggregator provides a number of value-adding functions. The first is filtering and consolidation: by topic, by relevance, and by quality (Katz 2004).

Filtering and consolidation in today's TV industry is carried out by program schedulers, who use their understanding of the target audience of a particular channel to schedule a sequence of programs aimed to hold their attention. This is as true for general entertainment channels as for the more specialized channels covering areas such as news, sports, history, arts, or technology. Note that a single channel is itself an aggregator, but companies that function in this space seek to address wider audiences by further aggregating a number of channels, either in-house produced, bought-in, or a combination of the two. The BBC and Sky in the UK are well-known examples.

The music recording industry aggregates music tracks within artists/genres as CDs, or increasingly as a themed inventory in online stores. Publishers offer technical books, novels, magazines, and newspapers and struggle to figure the Internet angle for content dissemination. The film industry's price-discriminating distribution sequence of theaters and cinemas, hotels and airlines, DVD, pay and subscription TV, and free-to-air TV, and their concerns with an Internet distribution model are also well-known.

By a careful use of their editorial, scheduling, and quality-assessment skills, the content aggregator can create a consolidated experience for the customer who has stability and predictability. This can create the basis of an enduring relationship that can be monetized—the basis of brand identity and brand power.

A second function of the content aggregator is a reduction in transaction costs, the standard role of the middleman. The content provider does not have to deal with a myriad of platform providers (or they with him), or try to secure payment from an unbounded number of consumers. Likewise the consumers are not faced with the impossible choice of finding and establishing relationships with a dynamically shifting population of content providers: the task is subcontracted. These are the forces that shore up the role of aggregator and resist disintermediation.

Internet Search Engines as Automated Content Aggregators?

But perhaps in the new world of the Internet, search engines can aggregate dynamically with less cost and overhead than existing people-intensive companies? Could Google be a content aggregator? A search engine has to solve the same set of problems as existing aggregators do, namely:

- Identify the relevant population of content providers,

- Rank them by topic, relevance and quality,
- Present them to an audience and permit selection,
- Ensure content delivery,
- Organize payment.

In principle, technical solutions exist to each of these problems. The issue is more how good these solutions are from the point of view of the customer experience. How would you rate Google News—automatically put together—against an online newspaper?

The relevant population of content providers is anyone publishing material to the Internet. This is not just as-yet-undiscovered bands. With conditional access systems enforcing digital rights management, digital material can be safely distributed over the Internet as we have seen with music download sites. Regulation is increasingly forcing owners of digital content to publish, for a fee, over the Internet. It seems that as more high-quality items finds themselves on the Internet, search engines should be able to locate them, and it is in the interests of the content publishers that they should do so.

Existing aggregators identify and rank talent by using specialist talent scouts—everyone has heard of the music industry's A&R men. Amazon's search engine ranks books both by sales and by customer reviews, both examples of user quality-assessment. Google ranks through a complex weighted page-link algorithm as a proxy for quality. Recommender systems matching your personal buying history with the buying patterns of similar customers have had some success. It would be unwise to bet that a search engine couldn't create a personalized menu of highly-valued content, whether TV, music, or textual material (a personal newspaper).

Once a search engine has identified and ranked a collection of possible content of interest, the result has to be presented to the user for final selection. There is a trade-off between the size of the selection problem you present to the consumer, and the sophistication and excellence of the filtering procedure that pre-selects what the customer would have selected anyway. Examples today include the Electronic Program Guides (EPGs) seen with multi-channel TV, listing guides in newspapers and specialist magazine, the screen of results returned by a search engine, and the themed catalogues seen on music download sites. Most Internet-hosted content distribution sites already contain an embedded search engine.

None of these existing formats should be seen as restrictive of the future. If there are consumers who simply want a personalized linear TV schedule for the evening, it will not be beyond the abilities of the search engine's programmers to design a system to put that together, if the primary content is out there at affordable prices in the first place. Skilled schedulers may believe that their jobs cannot be automated out of existence, but history does not appear to be on their side. A possible source of friction standing in Google et al.'s way is probably exclusive rights restrictions, which makes some of the highest value content unavailable except through the rights holder's branded channel. However, from the rights holder's point of view, a search engine is just another channel to market, so why not allow it to index your material: if an end-user then chooses it, it's additional revenue. The equilibrium probably makes a great deal of high-value content available to search engines.

Home consumption of Internet content can be done today from a PC via a broadband connection, and from a TV set via a broadband Set-Top Box (STB). With the arrival of higher-speed DSL or fiber links into the home, and a generation of usable media centers, IPTV and VoD from the Internet will be just another way to get TV and radio.

And, finally, there's content payment. This is far from being a trivial issue. No consumer is going to actively manage tens of micro-payments for an evening's viewing. And if prices for different content-items are all over the place, then selecting a satisfactory program schedule within a fixed budget for the evening is another complex optimization process that may or may not be able to be automated away. Content aggregators like Sky, the BBC, and other mainstream broadcasters solve the problem today by intelligent, experience-based scheduling for the mass audience that reliably delivers the number of viewers and commensurately rewards both advertisers and content providers whilst charging viewers a competitive and predictable fee. The search engine solution, with its radical customization of the schedule down to individual preferences, naturally finds pricing and billing more difficult, while promising increased satisfaction. Of course, family viewing is either shredded or the subject of complex negotiations in this model, another example of where you might have too much freedom?

Platform Providers

Originally there was only terrestrial transmission of TV and radio, and with a fixed and limited amount of spectrum, the result was channel scarcity. The spectrum generated revenue as long as it was used, so content aggregation into linear channels was an efficient response. The transmission bottleneck led to vertically-integrated companies that produced content, aggregated and scheduled it, and then broadcast it.

Cable and satellite platforms massively increased the number of channels but the basic nature of content distribution—one-to-many—still meant that customers could not be individually targeted. The industry was still about the competition between broadcasters for aggregate shares of the mass market and the revenues that were tied to audience share.

The Internet, with its point-to-point architecture, completely removes the limitation on number of channels (Katz 2004). In theory, every single user could have his own personal channel. At this level of granularity, the very concept of channel, as a prepacked linear ensemble of programs, loses its force. Since any content is in principle accessible at any time to any person, the architecture is *random* rather than *serial* access. The most general service provided is video-on-demand. Linear channels *may* be offered for commercial reasons, but they are not mandated by the technology, and some people may elect not to consume their entertainment that way at all.

The often heard argument that capacity does not exist on the Internet to provide unbounded choice is not compelling. A standard definition TV program needs 3–4 Mbps if you include metadata such as EPG updates as well. One wavelength on a fiber could easily carry 10 Gbps that would support 2,500 TV channels. It is not very hard to pipe one wavelength around a national optical network.

For VoD, the right architecture is local caching. One Terabyte disk arrays are just about consumer items—at time of writing—costing around $700. Robust carrier storage would be more expensive, but a server farm of 100 of these could store around 20,000 TV programs of two hours each. Seems a reasonable choice. In fact, the ideal architecture is to have a lot of program storage on the customer's own premises equipment based on their preference profiles, refreshed by background updates from larger cache server farms at the local network Point of Presence, with a centralized network archive server of last resort.

Finally, with 3G cellular and TV transmission networks (eventually to be joined by pervasive WiFi and WiMAX networks), and the current generation of multimedia handsets, it is now possible to receive TV on mobile devices as well. Mobile networks are not the Internet—the radio access network, which is shared, two-way and expensive, is not efficiently used for scheduled, nonpersonalized channels, even with MBMS (Multimedia Broadcast Multicast Service—the multicast service on 3G mobile networks). It is better to broadcast mobile linear TV on an overlay wireless network, and reserve the scarce capacity of the two-way radio access network for video-on-demand.

The Impact of the Internet

As we discussed previously, the conventional broadcast technology base for TV content distribution (satellite, cable, terrestrial) has three salient characteristics.

- It is one-way (with at best a nonintegrated and low-capacity PSTN back-channel).
- It is one-to-many (which means individual customers cannot be addressed, only aggregates).
- It is spectrum limited (which means there is a distribution bottleneck restricting supply).

A fourth characteristic is also cited—existing platforms are geographically limited, while the Internet is global in coverage and without boundaries. Geographical limitations are the basis of a whole content resale industry through syndication. For example, Warner Bros. sold new episodes of *Friends* to NBC for about $4 million an episode. They also sold reruns of the same episodes to hundreds of U.S. local stations for another $4 million per episode. Such geographically-based resale apparently generated more than $1 billion in syndication fees for the studio. If anyone in the world could watch *Friends* over the Internet, then this whole syndication edifice would come crashing down. But the technological hurdles to geographical localization on the Internet have been much exaggerated. A combination of IP address monitoring and user authentication via credit card (which can serve to check the user's address) can allow the Service Provider to restrict access perfectly well within geographical boundaries in accordance with the rights they have acquired. There will always be the possibilities of fraud at the margins, but the business model *can* be made robust.

The three characteristics of current platforms itemized above have shaped the industry. With only one-way, one-to-many platforms, programs have to be linearized into channels. This mandates the role of the content aggregator, who assembled and owned channels, as we have seen. With spectrum scarcity, only a restricted number of channels could be broadcast, and consequently per-channel viewing figures could be relatively high. This attracted advertising finance and underpinned a business model where power accrued to the aggregator.

Broadcasting consequently evolved into the standard market structure described in chapter 6, that of the "Rule of Three." In the United States, there are generalist stations such as ABC, NBC, CBS that were the traditional major players, and then a host of niche channels came along. Fox has had recent success as a new generalist, perhaps a pointer to changing times.

In the UK, there were historically five main free-to-air channels: BBC-1, BBC-2, ITV, Channel 4 and Channel 5. Historically BBC-1 and ITV were the generalist channels, with the remaining three more niche. The UK situation is dominated by the license-fee funded BBC, however.

The recent rise to prominence of satellite companies such as Sky and DirecTV, and cable companies such as NTL in the UK, Comcast, Time Warner, and Cox in the United States has fragmented the market somewhat, with many new channels, but has not dented the power of the content aggregators. The cable companies are geographic monopolies in the regions they serve, and compete with terrestrial free-to-air channels through premium content. The satellite providers also tend to be geographical monopolies—even a *duopoly* can cut returns significantly, satellite TV being a scale business.

Today, a consumer typically has a choice of signing up with one cable company, with one satellite company, or with a terrestrial broadcast platform. How will the new Internet and mobile platforms change this?

The Race to the Middle

In value-net terms, the pre-Internet value net was firmly under the control of the content aggregators. They owned the channels, they were the branded companies, they viewed their distribution platform as a utility, and they billed the customer/advertiser. With the arrival of a TV-capable Internet, new players arrived, the carriers, who were not necessarily resigned to a subordinate role in carrying bits for the major broadcasters. We are currently seeing, therefore, a race to the middle to establish control of the most valuable territory in the new value network (Figure 15.2).

In this diagram, the solid arrows show the points of departure. On the left-hand side, the broadcasters produce or acquire content, and ingest it in their head-ends. In some cases, the channels are then encrypted if they are part of a premium package or have any restrictions on usage rights The broadcasters also produce schedules for their own channel programming designed to maximize audience share. The distribution is over dedicated terrestrial radio, satellite, or cable networks, usually operated by an independent company, and on which the broadcaster has bought capacity (not shown in the diagram). In the case of satellite, cable, and digital terrestrial transmission, a set-top

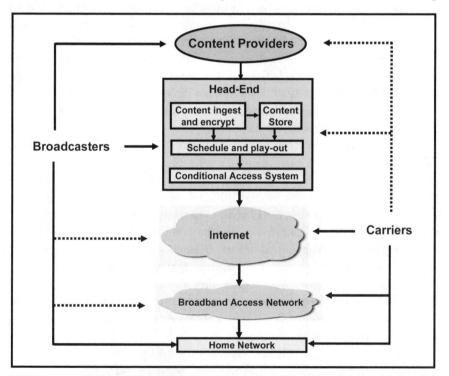

Figure 15.2 Internet platform: the race to the middle

box and possibly a viewing card is required to authenticate the user, confirm billing status, and help decrypts the signal. The user cannot normally access the distribution platform without such a device. Customers accessing purely free-to-air channels may be able to get by without the viewing card and point-of-use billing mechanisms.

The starting situation for the carriers is completely complementary, as shown by the solid arrows on the right hand side of the diagram. In the general case, the carrier owns the IP transmission network and the broadband access network, and provides or supports a broadband modem or router as the gateway device for the home network. The utility of the broadband connection to the customer is the access it gives to all the services available over the public Internet, very few of which are provided by the carrier itself. So for the carrier, this is a "pipes" business, one where only utility rates of return can be expected in a competitive or strongly regulated market.

Once it became technically possible to run TV content over the Internet and into people's homes via a broadband connection, the broadcast and Internet value chains became glued together, as shown in the diagram, and the major players from both camps began to circle each other warily, trying to determine where the value really was, and whether they were in a position to colonize the high-value portions, and extract economic rents.

The Broadcaster View

The Internet and mobile networks are both a new distribution platform for linear TV channels (IPTV) and the basis for a new network service—VoD. Neither service is technically hard to realize. IPTV and VoD require re-engineered head-ends, a transmission and broadband access network and a re-engineered STB.

The main issue facing broadcasters is whether to use the generic Internet as a platform, whether to resell a carrier's network offerings, or whether to forward integrate into the carrier space by acquiring or building their own IP networks.

It is likely that broadcasters will make their content available over the public Internet anyway—the incremental costs to do so are small and the revenue opportunities appear to be there via DRM. Regulation may, in any event, call for it. The additional benefits of forward integration into network ownership lie in the control over quality in these early days of the technology, and the ability to evade the kinds of "hold-up" we discussed in chapter 13, whereby carriers can exploit their network dominance to extract monopoly rents from upstream content providers.

An unintended consequence of forward integration is that the broadcaster becomes an IP service provider, with a portfolio of communications and Internet access services. The broadcasters are inclined to write these off as marginal products as compared to their very profitable content-based services. In the longer term, they may be wrong about this (cf. below and Odlyzko 2002).

The Carrier View

Most facilities-based carriers are today considering introducing IPTV and VoD services. They see these services as very profitable, and as substituting for their declining voice revenues. In terms of Figure 15.2, they would like to backwards integrate into the areas shown by dotted lines on the right-hand side.

Architecturally, the next-generation network is of great help. The NGN comes with QoS, call admission control and bandwidth management capabilities courtesy of the transport network and the IP Multimedia Subsystem. IMS additionally provides a session management and billing service that could easily be used to support a VOD service. Specifically, IMS supports the following relevant functions.

- User authentication via the HSS,
- User service profile management via the HSS,
- Billing based on the user service profile,
- Bandwidth allocation via the Go interface between P-CSCF and the first service routing device (PE or GGSN),
- Session admission control (via many components, most notably the P-CSCF),
- VoD server functionality via IMS Application Servers.

On this basis, there is no question of the adequacy of IMS to manage and bill a VoD service. However, IMS does not today directly support the encryption and key management functions of a Conditional Access (CA) system. Unfortunately, existing CA systems are quite tightly coupled to user authentication and billing, so unless CA systems are re-engineered in a modular fashion to interwork with IMS, this appears to be a significant roadblock. But well-within the capabilities of a carrier to resolve.

The broadcasters, of course, already have systems that do many of the middleware functions that IMS abstracts and modularizes. This makes IMS much less attractive to them. However, if broadcasters start to think more like carriers, and focus on the quadruple play of (video)-telephony + data + IPTV/VoD + mobile, then the increased generality of IMS may well justify deployment in a few years time.

BT, for example, has proposed to offer content management and distribution services to a wide variety of content owners, and has floated the idea of a home device—a hybrid Freeview box (free-to-air digital terrestrial TV channels) plus a

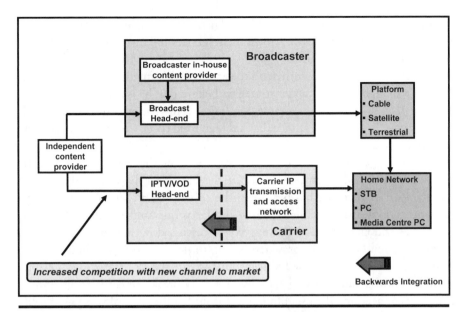

Figure 15.3   Carrier backward integration opens new opportunities.

broadband connection for VoD. It has also deprecated suggestions that it might get into content aggregation itself, in competition with the BBC, Sky, and others. At the very least, BT's initiative has the potential to disintermediate the existing content aggregators vis-à-vis content producers, and opens a competitive space for these and new entrant aggregators and portals to bring TV material to the mass market (Figure 15.3).

The carriers can support new content and media services over their new IP networks, and they can provide the infrastructure to adapt content onto them as well. They probably don't want to backwards integrate into either content production (buy a studio or production company) or content aggregation (buy or set-up channels), both of which require skills and culture at variance with those of carriers. All-in-all, this is not a bad strategy for the carriers to adopt.

Business Strategies in the Consumer Communications Services Sector

In 2001, Andrew Odlyzko wrote an interesting paper called "Content is not King." This argued that despite appearances, content provided neither the margins nor the revenues of communication services, and that this pattern would be repeated on the Internet. He predicted that 3G mobile networks, widely touted as the key enabler for new content services, would instead see most utility as a lower-cost, higher-capacity platform for voice calls. He presented plenty of evidence that amongst both business and residential users, there is a much greater demand for communication services than for content services and that this would continue to drive the revenue disparity.

As voice revenues apparently tend to zero, a skeptic might think Odlyzko has this all wrong, but many carrier strategists believe that he is absolutely right.

First of all, why are voice revenues tending to zero? The provision of voice calls is an extreme example of a service that has high fixed costs, which are infrequently incurred and are sunk, and almost zero short-run marginal costs up to the capacity limitations of the network. In the circuit-switched world, regulation of the wholesale price and schemes, such as Carrier Pre-Select, have created a highly-competitive market that has bid down the cost of voice calls to levels perhaps lower than long-run incremental cost—insofar as this can be estimated.

The second factor is VoIP. Putting Skype (discussed in chapter 9) to one side, there are a number of VoIP providers selling services that offer an almost perfect substitute for circuit-switched voice. The costs to the VoIP provider are soft-switches, media and signaling gateways, a PSTN break-out connection, and an Internet connection. The user often pays for the handset, and always pays for the broadband connection: the Internet itself is free from the point of view of both the user and the VoIP supplier. On this dramatically lowered cost base it is possible to undercut even competitively-priced circuit-switched voice, and the VoIP supplier is paid for incoming calls by the PSTN carrier. A Skype-like service, of course, saves on the soft-switches and associated operational costs.

So why are carriers unreasonably optimistic? People like to communicate, and every time the technology advances to a threshold of usability, a new service opportunity beckons. SMS has to be at the margins of usability, but is nevertheless a huge business. Without trying to predict the details of "new wave" communications services, it seems likely that a fully multi-media and pervasive transport network, overlaid with a sophisticated session management and charging mechanism, and combined with some new handset and terminal ideas, could launch a number of innovations.

For example, sometimes services languish for long periods because they are just not good enough, and then suddenly the technology improves and they take off. Mobile phones were clearly in that category, but looking ahead, I would guess that video-conferencing might develop in that direction too. As a frequent user a few years ago, I can testify that sound quality was often poor, the video cramped and inflexible, and the set-up and control interfaces opaque and barely usable. The recent popularity of HP's Halo system (http://www.hp.com/halo), which spares no expense to create a high-resolution sense of copresence, seems to suggest we are only just about at the point of doing this right. If videoconferencing works well for users follows the usual trend, then prices will come down and usage will explode over the next decade.

Even in voice services, there are opportunities for improved (CD) quality at enhanced price points. It's a mistake to look at voice only through the inflexible blinkers of circuit-switched telephony.

User-Generated Content

A further area of interest is user-generated content. Photos and video are the paradigmatic examples. With digital cameras, camcorders, and camera phones, what do people do? They take pictures and videos, select the best, and mail them to their friends or post them to specialized share-Web sites (e.g., http://www. youtube.com). There has been a relatively slow-take up of picture messaging on 3G phones, but this is probably due to a combination of early-adopter premium pricing, lack of usability, and the current low take-up of 3G handsets. There is no reason why picture and video messaging shouldn't be huge once the usual tipping point has been reached.

User-generated content is in the overlap area between content and communication services. That overlap space is richer than some people imagine. It is tempting to think of music, photo, and video share-sites as the exclusive preserve of enthusiastic amateurs. This is far from the truth. Even a cursory review of the more high-profile sites will show the prevalence of:

- New kinds of advertising, often rather edgy;
- Political and social commentary, clearly put together by funded interest groups;
- Promos, out-takes, and other spin-offs from established media archives.

It's clear that these sites are being used as a laboratory for many economic, social, corporate, religious, and political groups across the world. The production values are often higher than would be expected and the content is more accessed, and gets higher ratings, than the totally amateur material. As always, such pluralism and diversity is to be welcomed, and we already see hints on future mechanisms of monetisation (e.g., intrinsically interesting short ads).

Conclusions

The consumer market is a market for both the consumption of professionally produced content and for communications services that allow people to communicate with each other.

In the former case, the impact of the NGN presents a new platform opportunity to existing broadcasters, one they are eager to exploit: it adds interactivity and is the basis for new services such as video-on-demand. The carriers own the next-generation network, of course. They can sell access to the broadcasters, and by implementing generic head-end functions, they can partially move into the content aggregator space and cut separate deals with content providers. Broadcasters also have the option to forwards-integrate by acquiring and investing in alt-net operators with suitable networks. They may then have to come to terms with what they have bought.

In the latter case, person-to-person communications, the future looks good. Doomsayers point to the death of voice revenues, and discount future services to zero. The truth is wholly different: people have always tried to use new technologies to communicate with each other, even when the platforms are difficult to use. Get the pricing, performance, and usability right and usage explodes. We are still in the dark ages when it comes to the potential for technologies to enhance communications, so the future for carriers who can monetize this area is enormous.

References

Katz, M. L. 2004. Industry structure and competition absent distribution bottlenecks. In *Internet Television,* eds. E. Noam, J. Groebel, and D. Gerbarg, 31–59. Mahwah, NJ: Erlbaum., 2004.
Odlyzko, A. (2001). Content is not king. *First Monday*, Vol. 6, Number 2 (February 2001). http://www.firstmonday.org/issues/issue6_2/odlyzko/

Chapter 16

Conclusions

The story of the next-generation network is a history in three parts: the original carrier concept of Broadband-ISDN, the unlooked-for arrival of the Internet, and the current carrier project, the ITU-T standardized NGN.

Back in the 1980s, the telecoms industry was dominated by large, regulated monopolistic carriers. Technologies were often proprietary, always complex and innovation was refracted through cumbersome national, regional, and global standards bodies. Carriers knew that the technology of end-systems was rapidly evolving, with the invention of personal computers and LANs. They also knew that the future network would have to carry a combination of voice, video, and data. The project to move the voice-centric carrier networks to this multimedia future was called Broadband ISDN and was years in the making.

The major feature of carrier standards-setting is that the standards will define the equipment, and the equipment will be expensive and will be in the network for decades. Therefore, the standards have to anticipate the services for the next 10–15 years, and they end up by being very complex. And it all takes a very long time.

The main products in service in the 1980s were transmission and the public switched telephone network. Transmission is all about getting bits sent from place to place reliably, and this had been achieved in the 1980s with the PDH asynchronous digital transmission networks. Indeed these were on the point of being replaced by the more modern SONET/SDH networks. The PSTN was available in all offices and most households, and worked to a high degree of reliability.

It turned out that by simply adding two pieces of equipment, the existing PSTN and transmission networks across the world could be turned into a data network accessible to anyone with a phone. The two pieces of equipment were the modem and the router. Thus was born the Internet as a mass phenomenon, and the Internet Service Provider as the company that could package Internet access to customers.

ISPs constituted a classic competitive market. Entry was easy and the service was not highly differentiated (access, e-mail, Web hosting). Many value-added services were available, as PCs and servers, functioning as Internet end-systems, became more powerful: games, Internet shopping, information retrieval, and so on.

In the early years, carriers viewed the Internet as a frivolous distraction. However, as the nineties progressed, most carriers developed an Internet arm that often achieved a dominant market presence. As the Internet became carrier-grade, with new carrier routers and broadband access at the turn of the millennium, it gradually became clear that the Broadband ISDN dream was dead. But the Internet was not a ready-to-hand replacement.

The problem with the Internet was that it grew organically and incrementally, without any strategy. This was not good for companies that needed to plan ahead and to manage enormous capital budgets. The Internet was *protocol*-centric, developing new protocols as need became apparent, on-demand. The carriers, however, were *architecture*-centric, developing platforms to address both present and anticipated future needs, *especially* including billing, and that would need to be in service for an extended period.

It took a number of years for the Internet to catch up with carrier needs—specifically in the areas of multimedia signaling (SIP), security (IPsec, Diameter), and service differentiation (Diffserv). In the end, the carrier NGN architecture program was actually applying a *forcing function* to the IETF to develop its protocol suite to the level the carriers needed.

Over the next years (to around 2012) most of the work will be completed, and the Internet will evolve to the architecture of Figure 16.1. The next-generation network architecture retains the open, layered Internet architecture. There is no going back to the old days of closed, monolithic, and proprietary architectures. The other feature of note is that the carriers do not have a monopoly of supply at any layer. Sure, you can access a carrier application (D) running on a carrier application-hosting platform (F) with carrier voice, video, e-mail, and instant-messaging services (H) running on a carrier network (K). Everything will be integrated and, no doubt, there will be just one bill.

But you could equally imagine accessing a user-generated content site such as eBay (A) or an independent application services provider (C), running on an independent hosting company's platform (E), bundled with Skype (G), and running on a network infrastructure established by Google (J). At the price of paying several bills. And, no doubt in the UK, you will be watching Sky TV (B) over Sky's broadcasting/VoD system (E) over Sky's IP/MPLS network (J?, K?) and be back to one bill again. Does this make Sky a carrier?

Figure 16.1 indicates the main *functional* components that will be put in place over the next few years. Imagine each box (A–K) to be a country, capable

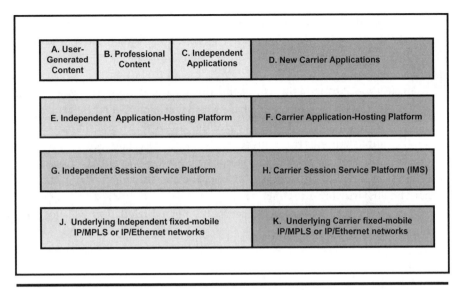

Figure 16.1   The next-generation architecture.

of being colored to represent it being the property of one or another empires. Then a value-net analysis identifies players such as carriers, broadcasters, Service Providers (Google, Yahoo!), and more generic E-businesses (e.g., eBay) and asks which boxes they own now, and which ones they aspire to own through forwards or backwards integration. Or is the trend the reverse, that the current multi-box ownership of vertical integration will fragment into independent companies in a process of dis-aggregation?

We had this discussion in chapter 15, where we saw examples of carriers moving into the content-aggregation and broadcasting space, while broadcasters acquired carriers and moved into session, hosting, and network services. There is every reason to believe that this process will work its way to completion over the next few years. We will therefore see large, vertically integrated companies from these very different backgrounds confronting each other in the new converged marketplace. However, I emphasize that the independents and innovators will still be there. Just as in all the creative arts, their challenge will be to get noticed, to get mass audience share, and then to avoid being suffocated by the embrace and extend strategy of the majors with their copycat acts, productions, applications, services, and Web sites: we have been here before with software. Or else the carriers will extract a large part of the value through rents as described in chapter 13.

So in conclusion, if we were looking back, say, from 2012, we would observe that the period 2007–2012 was when the Internet shed its bottom-up wild west image and became a sophisticated global utility that finally began to fulfill its service delivery potential. We would note that the carriers were still making a very good business running a variety of platform services, but that content companies had also successfully colonized this new distribution and communication medium. The degree of innovation would be greater, not less than today. The genie of user-generated content and innovative applications and services would not have been rebottled.

To make a specific prediction, I imagine that technical innovation in terminal devices together with ever-increasing bandwidth will allow something truly spectacular to arise in the areas of computer gaming (and virtual worlds in particular), and in the various spin-offs from this genre. Networked gaming seems to combine the best in professionally developed content with mass user-creativity in a scalable medium, with interactivity designed-in from the outset. Perhaps I'm just saying that it's finally virtual reality time!

And the need for intelligent and precisely targeted regulation will not have gone away.

Glossary

.NET   Microsoft's application development platform with an orientation to web services.

21CN   Twenty First Century Network—BT's program to migrate to an NGN and retire its current circuit-switched network in the UK.

2G   Second Generation mobile phone networks such as GSM. Typically used to carry voice and low-rate data.

3G   Third Generation mobile phone networks capable of supporting high data rates and focused on multi-media services.

3GPP   Third Generation Partnership project—consortium to develop a standards roadmap for 3G mobile services and technology evolving from GSM.

3GPP2   North American organization covering similar areas to 3GPP, starting from the CDMA technology base used in North America.

A&R   Artist and Repertoire—the division of a recording company responsible for scouting for and developing new talent.

AAA   Authentication, Authorization Accounting—functions carried out by IETF protocols and servers such as RADIUS and DIAMETER.

AES   Advanced Encryption Standard—current state-of-the-art in symmetric key encryption algorithms. The U.S. government standard.

AI   Artificial Intelligence—branch of computer science studying computational approaches to understanding and modeling psychological processes.

AIN   Advanced Intelligent Network—name used in North America for the circuit-switched Intelligent Network (IN).

AIP   Application Infrastructure Provider—a business that provides a platform for Internet-accessible applications—the customer loads their application on to the AIP platform. Relevant technologies: Java EE, .NET.

AJAX   Asynchronous JavaScript And XML—a technology to make the browser smarter by loading/executing Javascript and asynchronously fetching data from the server (using XML).

ANSI   American National Standards Institute.

AS   Application Server in IMS.

ASP   Application Service Provider—a business that provides software as a service over the Internet, accessed via browser. Salesforce.com is a well-known example. Relevant technologies: Java EE, .NET, AJAX.

ATM   Asynchronous Transfer Mode—the B-ISDN cell format for carrying all services across the network.

BGP   Border Gateway Protocol—Internet routing protocol.

BGP-4   The current version of BGP in service on the Internet.

B-ISDN   Broadband ISDN—the pre-Internet carrier vision of the multi-service network of the future.

BPE   Business Process Re-engineering—the discipline of analyzing, redesigning, and transforming business processes.

BRAS   Broadband Remote Access Server—network device which aggregates traffic from DSLAMs and terminates DSL sessions (e.g., PPP), and manages QoS and accounting functions.

BSC   Base Station Controller in GSM.

BSS   Base Station Subsystem in GSM.

BSS   Business Support Systems—IT used to run the business (e.g., CRM, ERM, billing, etc.).

BT   British Telecom—major carrier in the UK.

BTS   Base Transceiver Station in GSM.

CA   Conditional Access System—digital rights management system used to control access to content.

CAPEX   Capital Expenditure—that portion of an organization's budget which covers capital costs (e.g., new equipment).

CDR   Call Detail Record—used in billing.

CE   Customer Edge—the customer edge router which connects to the Service Provider network.

Centrex   The name of a carrier service offering PBX facilities to feature-rich handsets. Equivalent to a hosted PBX service.

CIO   Chief Information Officer—the executive in charge of Information Technology within an enterprise.

CORBA   Common Object Request Broker Architecture—a standard for distributed computing.

CoS   Class of Service—a network mechanism to divide traffic into classes for QoS-related handling.

COTS   Commercial Off-The-Shelf—systems that are bought and configured (vs. in-house developed applications).

CPS   Carrier Pre-Select—a pro-competitive arrangement where you can register your preferred telephony provider, and the incumbent is obligated to hand your call straight to that provider (replacing prefix digit dialing).

CRM   Customer Relationship Management—systems which store customer data, take orders, manage sales etc.

CSCF   Call Session Control Function—a component of IMS. An updated version of the soft-switch.

CTO   Chief Technology Officer—the executive in charge of Technology within an enterprise.

CUTV   Catch-Up TV—a form of VoD covering recently broadcast material.

CWDM   Coarse Wave Division Multiplexing—the provision of a number of optical channels via closely frequency-spaced laser carriers. Cheaper but fewer channels than DWDM.

DDoS   Distributed Denial of Service—a DoS attack launched from many different computers (e.g., by synchronously activating trojan programs previously covertly installed on each computer).

DHCP   Dynamic Host Configuration Protocol—assigns IP address and standard services to hosts as they power up on the network.

Diffserv   Differentiated Services—a method of implementing services classes on the Internet by marking the required class of service in a field in the IP packet header called the DSCP (Diffserv Code Point).

DNS   Domain Name System—the Internet distributed directory which maps names (e.g., URLs) to IP addresses.

DoS   Denial of Service—an attack on a Web site by bombarding it with protocol messages for the purpose of overloading it.

DRM   Digital Rights Management.

DSCP   Diffserv Code Point—a field in the IP packet header where the required class of service can be indicated.

DSL   Digital Subscriber Line—the technology to provide a digital service currently up to 24 Mbps down a copper pair phone line.

DSLAM   Digital Subscriber Line Access Multiplexor—the exchange equipment that connects to the phone line to provide DSL.

DT   Deutsche Telekom—major carrier in Germany.

DWDM   Dense Wave Division Multiplexing—providing large numbers of optical channels via closely frequency-spaced laser carriers.

EAI   Enterprise Application Integration—architecture and technology middleware for automatically linking applications.

ECM   Entitlement Control Message—used to distribute control words (decryption keys) in CA systems.

EMM   Entitlement Management Message—distributes authorization to decode to an STB in CA systems.

EPG   Electronic Program Guide—a directory used by the viewer to select content to watch.

ERM   Enterprise Resource Management—systems which support business functions (e.g., logistics, finance, HR, etc.).

ERP   Enterprise Resource Planning—a synonym for ERM.

eTOM   Enhanced Telecoms Operations Map—a standardized model of carrier business and operational processes.

ETSI   European Telecommunications Standards Institute.

FFM   Five Factor Model—the standard model of personality in contemporary academic research.

Frame Relay   Data protocol at layer 2 that replaced X.25. MPLS largely replaces it.

FT   France Telecom—major carrier in France.

GCF   Global Grid Forum—organizing body for grid computing.

GGSN   Gateway GPRS Support Node—router connecting the GPRS cellular access network to the chosen IP network (e.g., the Internet or an enterprise network).

GMPLS   Generalized MPLS—an extended form of MPLS used for managing virtual circuits in layer-2, TDM, optical and fiber networks.

GPRS   General Packet Radio Service in mobile telephony.

GSM   Global System for Mobile Communications—basis for European-developed 2G mobile phone system.

GTP   GPRS Tunneling Protocol in GPRS. The tunnels connect an SGSN to a GGSN.

GUID   Globally Unique Identifier—a document identifier used in Freenet.

H.323   A family of protocols originally designed (in the ITU-T) for LAN video-telephony. H.323 was initially pressed into service for VoIP implementations, but is increasingly being replaced by SIP and IMS.

HLR   Home Location Register in GSM.

HPP   High-Performance People—a Human Resources term for talent to be fast-tracked within a company.   HSS   Home Subscriber Server in IMS.

HTML   HyperText Markup Language—an annotation language for designing Web pages.

HTTP   HyperText Transfer Protocol—a protocol for exchanging messages between Web browsers and Web servers.

HTTPS   Refers to the layering of HTTP over SSL (e.g., to provide a secure Web site connection).

IBGP   Interior BGP—the variant of BGP used within a Service Provider's network.

I-CSCF   Interrogating Call Session Control Function in IMS.

ICT   Information and Communication Technologies—an acronym capturing the convergence between computing and telecommunications.

IETF   Internet Engineering Task Force—the global, voluntary community of engineers who develop and set Internet standards.

IMS   IP Multimedia Subsystem—the session signaling layer of the NGN.

IN   Intelligent Network—a circuit-switched public switching architecture that separates switching and services.

IP Internet Protocol—the packet format of data carried on today's Internet (IPv4).

IPsec IP Security—a protocol for encrypting traffic end-to-end over IP networks.

IPTV Linear TV channels distributed over an IP network.

IPv4 IP version 4—the current version of IP used in today's Internet.

IPv6 IP version 6—a version of IP which was aimed to replace IPv4, but which so far has found little acceptance.

ISDN Integrated Services Digital Network—the original carrier narrowband model of a multiservice network.

ISP Internet Service Provider.

ITU-T International Telecommunication Union Telecommunications Standardization Sector —part of the UN. Global carrier standards body.

J2EE Java 2 Platform Enterprise Edition—Java-based application development platform with a focus to Web services. See Java EE.

Java EE Java Platform, Enterprise Edition—the current name for the platform formerly known as J2EE.

L2TP Layer 2 Tunneling Protocol.

LAN Local Area Network—usually an Ethernet network which connects computers in an building.

LCAS Link Capacity Adjustment Scheme—used for managing the bandwidth provided by VCAT in next-generation SDH networks.

LRIC Long Run Incremental Cost—the pricing level regulators often set for companies with Significant Market Power. It approximates to the price which would be charged in a sustainable competitive market.

LSP Label-Switched Path—a "'virtual circuit" traversed by labeled packets in MPLS.

M&A Mergers and Acquisitions—a way of growing a business by buying or merging with others.

MAN Metropolitan Area Network—a network on the scale of a city or small region.

MBMS Multimedia Broadcast Multicast Service—multicast service on 3G mobile networks.

MBTI Myers-Briggs Type Indicator (TM)—a psychometric test classifying personality based on the theories of Isabel Myers and Katherine Cook Briggs (developing Carl Jung's theory of psychological types).

MDM Master Data Management—an IBM marketing term for managing an enterprise data model.

MG Media Gateway—a device that terminates circuit-switch traffic and maps it to/from IP packetization.

MGCF Media Gateway Control Function—a function within IMS which controls media and signaling gateways.

MNO   Mobile Network Operator—a mobile phone company with a network (contrasted with MVNO).

MP3   Audio compression format defined within MPEG-1.

MPEG-n   Moving Picture Experts Group—n = 1, 2, 3, 4 defines a set of protocols for compressed TV and audio.

MPLS   Multi-Protocol Label Switching—a labeling technology that allows virtual circuits to be created.

MSAN   MultiService Access Node—an enhanced DSLAM that also provides voice telephony (VoIP conversion), ISDN, and other data and transmission encapsulation services. It is typically a layer 2 edge device located at a carrier PoP.

MSC   Mobile Switching Centre in GSM.

MVNO   Mobile Virtual Network Operator—a "front-office" company that sells mobile services to customers and buys network capacity wholesale from another facilities-based operator, which owns a mobile network.

NASS   Network Attachment SubSystem—one of the subsystems in the NGN architecture. Used for IP address allocation, AAA functions and layer 3 location management (mobile IP).

NAT   Network Address Translation—typically used to map private IP addresses to public ones for Internet access.

NBMA   Non-Broadcast Multiple Access—denotes a full-mesh configuration of PVCs in Frame Relay.

NGN   Next-Generation Network—the carrier vision of an all-IP multiservice fixed-mobile converged network creating many new revenue streams and with dramatically decreased costs.

OCEAN   A Mnemonic for the five factors of academic personality theory: Openness, Conscientiousness, Extraversion, Agreeableness, Neuroticism.

OC-n   Optical Carrier—North American transport container (e.g., OC-3 = STM-1 = 155.52 Mbps).

OGSA   Open Grid Services Architecture—architecture standard for grid computing.

OPEX   Operational Expenditure—that portion of an organization's budget which covers running costs.

OSPF   Open Shortest Path First—an IETF routing protocol.

OSS   Operations Support Systems—in a carrier, the systems that manage the network itself.

OTN   Optical Transport Network—ITU standards covering transport, multiplexing, routing, management, supervision, and survivability of optical channels carrying higher layer traffic.

P   Provider—the Service Provider core routers which interconnect PE routers.

P&L   Profit and Loss—in business, having P&L responsibilities gives significant organizational power, and is therefore a goal of ambitious executives.

P2P   Peer-to-Peer—an architecture for applications which run on end-system, using the network just to carry traffic.

PBB   Provider Backbone Bridge—also known as MAC-in-MAC. A protocol being developed by the IEEE in 802.1ah. An additional header pre-pended to the customer Ethernet frame that allows carrier forwarding to be scalably decoupled from enterprise switching.

PBT   Provider Backbone Transport—a development of Ethernet that provides it with MPLS-like forwarding and resilience properties.

PBX   Private Branch Exchange—an almost meaningless acronym. It means a private telephone switch used to connect phone calls within an office or enterprise.

P-CSCF   Proxy Call Session Control Function in IMS.

PDA   Personal Digital Assistant.

PDH   Plesiochronous Digital Hierarchy—carrier transmission standard and technology— the technology is now obsolete.

PE   Provider Edge—the Service Provider edge router that connects to customer routers.

PMB   Project/Program Management Board—a review body in many program management methodologies.

PoC   Push-to-talk over Cellular—walkie-talkie function that will be an early mobile IMS service.

PoP   Point of Presence—the location of carrier equipment that is closest to, and directly connects to, the customer.

POTS   Plain Old Telephone Service—playful name for ordinary (legacy) circuit-switched telephony.

PPP   Point to Point Protocol—an IP packet framing, layer 2 protocol used for link error-checking, the carriage of connection authentication and password data, and for interface configuration. Commonly used on dial-up and DSL connections.

PRM   Project/Program Review Meeting—a regularly scheduled meeting in many program management methodologies.

PRM   Partner Relationship Management—a business function whereby an enterprise manages its relationship with its business partners (e.g., a carrier selling through VARs).

PSTN   Public Switched Telephone network—the current global circuit-switched phone network.

PVC   Permanent Virtual Circuit—in STM or Frame Relay.

PVR   Personal Vide Recorder—hard disk system for recording transmitted content for later replay.

QoS   Quality of Service—service quality as experienced by the user.

RACS   Resource and Admission Control Subsystem—one of the subsystems in the NGN. Used for admission control and bandwidth management.

RAN   Radio Access Network in mobile telephony.

RFC-n   Request For Comments (number n)—IETF standards document.

RFI   Request For Information—a tendering document issued to suppliers.

RFP   Request For Product—a tendering document issued to suppliers.

RFQ   Request For Quote—a tendering document issued to suppliers.

RFS   Ready for Service—the point at which a network, after construction, is handed across to operations to be enter normal service.

RIAA   Recording Industry Association of America—industry group active in legal action against Internet music copyright violators.

RIP   Routing Information Protocol—an older routing protocol.

RP   Rete Populi—"the people's network." A fictional technology in chapter 11 loosely modeled on Freenet.

RSA   Public Key Encryption algorithm invented by Ron Rivest, Adi Shamir, and Leonard Adleman in 1977. GCHQ had developed the basic technique in 1973, but this was kept secret.

RSVP   Resource Reservation Protocol—an IETF protocol that allows hosts to reserve session capacity across a network (with RSVP-aware routers). Generally felt not to scale on the Internet, where Diffserv is preferred (if any CoS solution is necessary).

SBC   Session Border Controller—a security gateway that sits in the voice and signaling path between VoIP network operators. It provides security and transcoding services.

SCM   Supply Chain Management—processes and systems whereby a business manages its suppliers.

SCP   Service Control Point—the services platform in an Intelligent Network.

S-CSCF   Serving Call Session Control Function in IMS.

SDH   Synchronous Digital Hierarchy—carrier transmission standard and technology—the basis for most current carrier networks.

SDH-NG   SDH Next Generation—the use of VCAT, LCAS standards to make SDH (and SONET) more data-capable.

SG   Signaling Gateway—a devices which maps circuit-switched signaling to/from signaling over IP.

SG&A   Sales, General and Administrative—accounting name for these overhead costs. Normally a line item in Profit and Loss accounts, or in a business case financial model.

SGSN   Serving GPRS Support Node—part of GPRS. The router connecting directly to the user terminal via the RAN.

SI   Systems Integrator—a company specializing in integrating complex components to deliver customer solutions. Broadly synonymous with high-end VAR, although they tend to prefer professional services or consultancy labels.

SIP   Session Initiation Protocol—an IETF protocol for managing multimedia calls on IP networks.

SIPPING   Joint IETF-3GPP working group to develop IETF protocols for the purposes of 3G networks (based on IP).

SLA   Service Level Agreement—a contract between Service Provider and customer specifying levels of service, penalty clauses, and so forth.

SME   Small and Medium Enterprises—a market segmentation category.

SMP   Significant Market Power—a term regulators use to identify companies that are candidates for regulatory attention.

SMS   Short Message Service—"texting." The 2G service that allows short messages to be sent between mobile phones.

SOAP   A protocol for exchanging XML messages between applications across a network - part of the Web services architecture.

SONET   Synchronous Optical Network—North American version of SDH.

SS7   Signaling System 7—the protocol used to set-up, manage and tear-down calls within existing PSTN networks (ISDN or analogue signaling is used in the access network—the ' "ast mile").

SSL   Secure Sockets Layer—a transport layer en/decryption and authentication scheme frequently used by Web sites (see HTTPS).

SSP   Service Switching Point—the switching function in an Intelligent Network.

STB   Set-Top Box—the device that terminates the TV signal and adapts it to the TV set.

STM-n   Synchronous Transport Module—a container for carrying data over a carrier transmission network. STM-1 = OC-3 = 155.52 Mbps.

SVP   Senior Vice President (senior executive title).

TCP   Transmission Control Protocol—used for reliably transferring files between end points. A layer-4 protocol layered over IP.

TCP/IP   See TCP and IP.

TDM   Time-Division Multiplexing—technique of using a high-speed bit-stream to carry interleaved voice samples from different conversations on the same wire or fiber in a standardized frame structure (PDH or SDH) in today's circuit-switched networks.

TISPAN   Telecoms & Internet converged Services & Protocols for Advanced Networks—ETSI NGN working group.

UDDI   Universal Description, Discovery and Integration—the application interface directory in Web services.

UDP   User Datagram Protocol—the other layer-4 protocol to TCP used to encapsulate data in IP packets. Used for low-overhead data connections where automatic retransmission-on-failure is not required (e.g., DNS queries, VoIP calls).

UMA   Unlicensed Mobile Access—GSM over WiFi scheme.

UNC   UMA Network Controller—the device used to translate voice and signaling from standard GSM format into the special format for carriage over an IP and DSL network.

URL   Uniform Resource Locator—the name for a networked informational resource (e.g., a Web site name, www...) .

VAR   Value Added Reseller—a company that typically integrates and customizes commodity products to produce a package to solve a customer problem.

VCAT   Virtual Concatenation—method in SDH-NG to create higher-capacity data pipes by bonding VC-n containers together (see LCAS).

VC-n   Virtual Container (e.g., VC-12, VC-3, VC-4). Container for data within SONET/SDH transport streams.

VDSL   Very high data rate Digital Subscriber Line—technology for data rates around 100 Mbps on copper wiring of a few hundred meters.

VLR   Visitor Location Register in GSM.

VNO   Virtual Network Operator—a "front office" company that sells communications services to customers, but buys network capacity wholesale from a facilities-based operator owning a network.

VoD   Video on Demand—technique/service of storing a number of TV programs on a server and allowing any program to be accessed by a user when that user so desires (on demand).

VoIP   Voice over IP—carrying digital voice samples (often compressed) across the network within an IP packet. Contrasted with circuit-switched time-division multiplexed carriage of voice, or carrying voice in other packet formats such as ATM.

VPN   Virtual Private Network—a service emulating dedicated inter-site transmission links on a shared network.

VRF   Virtual Routing and Forwarding instance (virtual router) within a PE device supporting BGP/MPLS VPNs.

W3C   World-Wide Web Consortium—standards body for Web technologies.

WAN   Wide-Area Network—often a carrier-provided network.

WiFi   Wireless LAN technology as specified by the family of protocols 802.11.

WiMAX   Worldwide Interoperability for Microwave Access—the family of wireless MAN protocols (802.16 and variants).

X.25   Obsolete data networking protocol and technology.

XML   eXtensible Markup Language—a syntactic framework for creating application-dependent markup languages.

Index

Page numbers in italics refer to figures or tables.